AF606528

18 0117865 X

# A DICTIONARY OF ELECTROCHEMISTRY

*Also published in this series:*

A DICTIONARY OF CHROMATOGRAPHY
R. C. Denney

A DICTIONARY OF SPECTROSCOPY
R. C. Denney

A DICTIONARY OF THERMODYNAMICS
A. M. James

# A DICTIONARY OF ELECTROCHEMISTRY

C. W. Davies

*D.Sc., C.Chem., F.R.I.C.*
*Professor Emeritus, University of Wales*

and

A. M. James

*M.A., D.Phil., D.Sc., C.Chem., F.R.I.C.*
*Professor of Physical Chemistry, Bedford College, University of London*

M

This book is sold subject to the standard conditions of the Net Book Agreement.

First published 1976 by

THE MACMILLAN PRESS LTD.

London and Basingstoke
Associated companies in New York Dublin
Melbourne Johannesburg and Madras

SBN 333 19203 6

Filmset at The Universities Press, Belfast, Northern Ireland and printed in Great Britain by Lowe & Brydone (Printers) Ltd, Thetford, Norfolk

# Preface

This work is a companion volume to *A Dictionary of Thermodynamics*, which is also in the Dictionary series, and it is intended as a handy reference book, both for students of chemistry and also for scientists in other disciplines, for whom it provides thumb-nail sketches of the important concepts of electrochemistry. The entries are in alphabetical sequence; this makes it easy to locate a given topic, and any disadvantage in this arrangement has been removed, we hope, by the provision of ample cross-references to relevant material. It is not intended that any topic should be treated rigorously; equations are quoted without derivation and their applications and limits are discussed. To all practising scientists this is a very important aspect.

It is written essentially for scientists who have a minimum of G.C.E. 'A'-level (or its equivalent) in a physical science subject. Besides chemists and chemical engineers, it is hoped that physicists, geologists and workers in the medical and biological sciences will find this a useful reference book, which in addition to definitions of the various quantities, etc., also contains tabulated data and provides access to standard reference texts for more detailed treatment. While not specifically aimed at schools, it is hoped that copies of the book will be available in the science libraries for more advanced students.

We are indebted to Dr. W. H. Lee of Surrey University for suggesting the project and for much practical help, advice and many useful discussions during the preparation of the manuscript. Our sincere thanks are due to our wives, for their understanding and patience during the writing of the book and to Dr. Mary P. Lord for her help and assistance during the preparation of the manuscript.

C.W.D.
A.M.J.

# How to Use this Dictionary

Entries are arranged in alphabetical sequence with full cross-referencing where there is more than one acceptable or recognised name.

Symbols and abbreviations are used in the text without definition; reference should be made to the list of Principal Symbols, p. vii.

In the text, words in italics followed by (q.v.) indicate a reference to another entry which would be of help; in some instances the reference to another entry is in parentheses. Thus in the sentence 'In an *electrolytic cell* (q.v.) electrical energy from an external source brings about a desired chemical reaction (see *electrodeposition of metals, electrolysis of water*)', the reader is referred to entries 'electrolytic cell', 'electrodeposition of metals' and 'electrolysis of water', for further information. Words in italics, followed by ‡, e.g. *free energy*‡, indicate a reference to that entry in the companion work *A Dictionary of Thermodynamics*, by A. M. James.

More detailed treatments of some of the entries can be found in standard reference books; where appropriate, relevant references are indicated at the end of an entry by a simple code for the name of the author. All the books, together with their codings, are listed in the Bibliography.

# Principal Symbols

| | |
|---|---|
| $A$ | relative atomic mass; area; Debye–Hückel constant |
| A | ampere |
| $B$ | Debye–Hückel constant |
| $C$ | capacitance; cell constant |
| C | coulomb |
| $D$ | diffusion coefficient |
| $E$ | electromotive force |
| $E(\mathrm{O, R})$, $E(\mathrm{X^+, X})$ | electrode potential |
| $E_\mathrm{A}$ | activation energy |
| $F$ | Faraday constant |
| F | farad |
| $G$ | Gibbs free energy function |
| $H$ | enthalpy or heat content |
| Hz | hertz |
| $I$ | ionic strength; electric current |
| J | joule |
| $K$ | equilibrium constant |
| $K_\mathrm{a}$, $K_\mathrm{b}$ | acid and base ionisation constants |
| $K_\mathrm{w}$ | ionisation constant of water |
| K | kelvin |
| N | newton |
| $N_\mathrm{A}$ | Avogadro constant |
| $P$ | total pressure of system |
| $Q$ | electric charge |
| $R$ | gas constant; resistance |
| $S$ | entropy; coefficient in Onsager's equation for conductance |
| $T$ | temperature/K |
| $V$ | volume |
| V | volt |
| $a$ | mean ionic diameter |

*Principal Symbols*

| | |
|---|---|
| $a_A$, $a_B$, $a_i$ | activity of A, B or *i*th component |
| $a_+$, $a_-$, $a_\pm$ | activity of cation, anion; mean ionic activity |
| $c_A$ or [A] | concentration of A/mol $dm^{-3}$ |
| $e$ | electron, electronic charge |
| $g$ | gravitational constant |
| g | gram |
| $j$ | current density |
| $j_0$ | exchange current density |
| $k$ | Boltzmann constant |
| $k_1$, $k_2$ | reaction rate constants |
| k | kilo (prefix), e.g. kg = kilogramme = $10^3$ g |
| $l$ | length |
| $m_A$ | molality of A |
| $m_+$, $m_-$, $m_\pm$ | molality of cation, anion; mean ionic molality |
| m | metre; milli (prefix), e.g. mm = millimetre = $10^{-3}$ m |
| $n$ | number of electrons transferred |
| $n_A$, $n_B$ | number of molecules of A, B in the system |
| $p$ | pressure above solution |
| $p_i$ | partial pressure of *i* in system |
| $s$ | solubility |
| s | second |
| $t$ | time |
| $t_+$, $t_-$ | transport number of cation, anion |
| $u$ | electric mobility of ion |
| $v$ | velocity |
| $w_A$, $w_B$ | weight of A, B |
| $x_A$, $x_B$ | mole fraction of A, B in solution |
| $z_A$, $z_B$ | charge number of ion A, B |
| $\alpha$ | degree of association or dissociation; transfer coefficient |
| $\beta$ | symmetry factor |
| $\gamma_i$, $\gamma_+$, $\gamma_-$, $\gamma_\pm$ | activity coefficient of *i*, cation, anion; mean ionic activity coefficient |
| $\delta$ | thickness of diffusion layer |
| $\varepsilon$ | permittivity |
| $\varepsilon_0$ | permittivity of a vacuum |
| $\varepsilon_r$ | relative permittivity |

| | |
|---|---|
| $\zeta$ | electrokinetic potential |
| $\eta$ | overpotential; overvoltage; viscosity |
| $\theta$ | fraction of adsorbent surface covered |
| $\kappa$ | electrolytic conductivity |
| $\mu_A$, $\mu_B$ | chemical potential of A, B |
| $\nu$, $\nu_+$, $\nu_-$ | number of ions, cations, anions formed from 1 mole of electrolyte |
| $\pi$ | ratio of circumference to diameter of circle $=3.141\ 59$ |
| $\rho$ | density; resistivity |
| $\sigma$ | surface charge density |
| $\phi$ | work function; potential difference at interface |
| $\Delta$ | increase in thermodynamic function, e.g. $\Delta X = X_2 - X_1$ |
| $\Lambda$ | molar conductivity |
| $\Omega$ | ohm |

**Superscripts**

| | |
|---|---|
| $\ominus$ | indicating a standard value of a property |
| $\infty$ | value of property at infinite dilution |

**Subscripts**

| | |
|---|---|
| A, B . . . | referring to substances A, B . . . |
| $i$ | referring to typical ionic species $i$ |
| $p$, $V$, $T$, $S$ | indicating constant pressure, volume, temperature, entropy |
| +, − | referring to positive, negative, ion |
| 1, 2 | referring to different systems or states of system |
| g, l, s | referring to gaseous, liquid or solid state, respectively |

**Other abbreviations**

| | |
|---|---|
| aq | in dilute aqueous solution |
| b.p. | boiling point |
| m.p. | melting point |
| v.p. | vapour pressure |
| pH | $-\log c(H^+)$ |

*Principal Symbols*

| | |
|---|---|
| p$K$ | $-\log K$ |
| [A] | concentration of A/mol $dm^{-3}$ |
| e.m.f. | electromotive force |
| $\approx$ | approximately equal to |
| $<$ | smaller than |
| $>$ | larger than |
| $\sum_i$ | sum of $i$ terms |
| $\prod_i$ | product of $i$ terms |
| $\exp(x)$ | exponential of $x$ |
| e | base of natural logarithms |
| $\ln x$ | natural logarithm of $x$ |
| $\log x$ | common logarithm of $x$ (to base 10) $\ln x = 2.303 \log x$ |

# A

## Accumulator

An accumulator is a voltaic cell which can be recharged by passing a current through the system from an external source. Many cell reactions that are chemically reversible prove unsuitable for this purpose because the charge–discharge cycle tends to bring about physical changes, e.g. in the condition of the electrodes. The practical choice is also limited to systems needing only a single electrolyte. Lead accumulators and the nickel–iron or nickel–cadmium cells are the most important, but the *silver–zinc cell* (q.v.), the *zinc–air cell* (q.v.) and the alkaline manganese cell (see *dry cell*) can also be used in storage batteries.

*The lead accumulator*

The reactions of the lead accumulator are

$$PbO_2 + Pb + 2H_2SO_4(aq) \underset{\text{charge}}{\overset{\text{discharge}}{\rightleftharpoons}} 2PbSO_4 + 2H_2O \qquad (A.1)$$

The electrodes are usually made of a grid of a lead–antimony alloy (which gives mechanical strength), filled with a paste of litharge and red lead in sulphuric acid. Fillers may be added to mitigate the disruptive effects of the large volume changes that accompany the cell reaction, and soluble substances to increase the porosity. When these plates are electrolysed in sulphuric acid, the charging reaction of equation (A.1) converts the anode material into porous lead dioxide, and the cathode material is reduced to spongy lead. Each cell of a lead storage battery usually contains a cathode consisting of a number of such plates connected in parallel, and, slotting into these, a similar arrangement of anode plates, each pair of plates being separated by a thin porous sheet of glass fibre or some plastic material.

The electrolyte is an aqueous solution of sulphuric acid. This is consumed in the discharge reaction, and enough must be present to provide for this and also for a sufficient excess to give good conductivity in the discharged cell. The more concentrated the acid used, the lighter will be the cell; but at very high concentrations the conductivity decreases and the freezing point rises. The optimum value is about 35% $H_2SO_4$ by weight, and the conductivity is then near its maximum throughout the charge–discharge cycle. The specific gravity is about 1.26, falling to about 1.1 when the cell is discharged. This gives a simple way of checking the state of charge of the cell.

The two electrode reactions have been studied under near-reversible conditions, and are

$$PbO_2 + 4H^+ + SO_4^{2-} + 2e \rightarrow PbSO_4 + 2H_2O$$

and

$$Pb + SO_4^{2-} \rightarrow PbSO_4 + 2e \qquad (A.2)$$

Equation (A.2) can be written

$$PbO_2 + 4H^+ \rightarrow Pb^{4+} + 2H_2O$$

$$Pb^{4+} + 2e \rightarrow Pb^{2+}$$

$$Pb^{2+} + SO_4^{2-} \rightarrow PbSO_4$$

The potential difference corresponding to the electrochemical step is

$$E = E^{\ominus}(Pb^{4+}, Pb^{2+}) + \frac{RT}{2F} \ln ([Pb^{4+}]/[Pb^{2+}])$$

Both concentrations are very small and nearly equal, so the reversible potential, +1.74 V, is practically that of the standard lead(IV)–lead(II) electrode.

At the other electrode the reactions

$$Pb \rightarrow Pb^{2+} + 2e$$

and

$$Pb^{2+} + SO_4^{2-} \rightarrow PbSO_4$$

give a reversible potential:

$$E = E^{\ominus}(Pb^{2+}, Pb) + 0.029 \log [Pb^{2+}] \qquad (A.3)$$

The concentration of lead ions in sulphuric acid saturated with lead sulphate is $5 \times 10^{-6}$ mol dm$^{-3}$, and equation (A.3) becomes

$$E = -0.126 - (0.029 \times 5.3) = -0.28 \text{ V}$$

The theoretical voltage is therefore 1.74 + 0.28 = 2.02 V, which is very close to the values found.

Typical curves for the charge and discharge of a lead accumulator

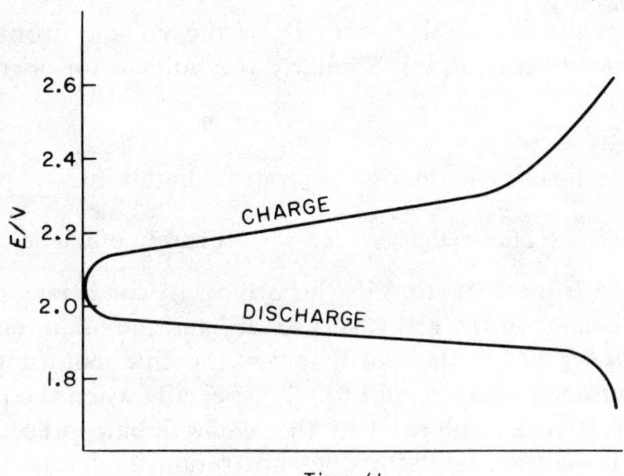


Figure A.1 Charge and discharge curves for a lead accumulator

are shown in figure A.1. Charging proceeds at an almost steady voltage after an initial rise, but eventually the voltage turns rapidly upwards and reaches the value at which hydrogen and oxygen are evolved freely. This marks the completion of charging, and a continued evolution of gas not only is a waste of energy but may assist the disintegration of the plates. During discharge the e.m.f. falls quickly to just below 2 V, and then gradually to 1.8 V. The cell should be recharged before the final rapid drop in voltage is reached.

The current efficiency, or ampere-hour efficiency, of a lead accumulator—(quantity of electricity given during discharge)/(quantity of electricity required for charging)—is higher than 90%. The energy efficiency, or watt-hour efficiency, is given by

$$\frac{\text{discharge voltage} \times \text{coulombs given}}{\text{charge voltage} \times \text{coulombs supplied}}$$

This is much lower because of the voltaic interval between the charge and discharge curves. The charging voltage is necessarily higher, and the discharge voltage lower, than the reversible e.m.f. of the cell. The voltage during discharge may be written

$$E_{\text{dis}} = E_{\text{rev}} - IR - \eta$$

where $E_{rev}$ is the reversible value, $IR$ is the voltage drop in the cell, and $\eta$ the *overvoltage* (q.v.). Similarly the voltage for charging is

$$E_{ch} = E_{rev} + IR + \eta$$

The energy efficiency is thus given approximately by

$$(E_{rev} - IR - \eta)/(E_{rev} + IR + \eta) \times \text{current efficiency}$$

Values range from 75% to 85% according to conditions of use. The internal resistance of the cell is very small, and the main reason for the loss in efficiency lies in the two $\eta$ terms; the chief contribution comes from *concentration overpotential* (q.v.), especially when the pores of the electrode are partially blocked by the lead sulphate precipitated.

A charged lead accumulator does not retain its charge indefinitely. The loss of charge can roughly be put at an average of 1% per day. The loss can be attributed to a variety of causes. At the positive plate lead dioxide can be lost by local action with the lead of the grid, and electrochemical corrosion will also occur at the site of any metal impurity in the lead electrode. To mimimise this danger, distilled water must be used in 'topping up' the electrolyte.

When a lead–acid cell is operating, lead sulphate is deposited in the pores and at the surfaces of the electrodes in a very finely divided form. On standing, this will tend to consolidate into larger crystals which can block the pores, thus reducing the electrode surface, or can be 'shed' from the electrodes and fall to the bottom of the cell. This is known as sulphating, and is the chief cause of the deterioration of cells. To minimise the effect, a cell should not be allowed to stand in a partially discharged state.

*Alkaline accumulators*

These cells operate on the reduction of tervalent to divalent nickel on one plate and the dissolution of iron or cadmium at the other. The electrolyte, which is not consumed in the cell reaction, is a 20% solution of potassium hydroxide, and addition of some lithium hydroxide is found to improve performance. The electrode reactions are

$$NiO(OH) + H_2O + e \rightarrow Ni(OH)_2 + OH^- \qquad (A.4)$$

and

$$Fe + 2OH^- \rightarrow Fe(OH)_2 + 2e \qquad (A.5)$$

and the over-all reaction is

$$2NiO(OH) + Fe + 2H_2O \rightarrow 2Ni(OH)_2 + Fe(OH)_2$$

In nickel–cadmium cells Cd replaces Fe in these equations. All the materials are insoluble, and the only change in the electrolyte is a slight increase in concentration as water is used in the reaction.

In the discharged state the active materials are nickel(II) and iron(II) hydroxides. These are poor conductors, so the $Ni(OH)_2$ is mixed with graphite or flakes of metallic nickel, and the mixture contained in pockets in a perforated steel plate to form the positive electrode. The negative plate is similarly filled with a finely divided mixture of iron(II) hydroxide and iron. A little mercury(II) oxide may be added, which, on reduction, gives a conductive film of mercury. The assembly can now be charged, when the reverse reactions of equations (A.4) and (A.5) take place. The oxidation of iron in the discharge process could be carried further, to the tervalent state, but this does not happen appreciably while the iron(II) hydroxide is precipitating, and it is avoided because the reduction of iron(III) oxide is not readily reversible.

Alkaline accumulators are much lighter than lead–acid cells, but give lower voltages. The working voltage of a nickel–iron cell is about 1.3–1.2 V, and for the cadmium cell it is somewhat smaller. The current efficiencies are less than for the lead accumulator and, as there is also a much wider gap between the charge and discharge voltages, the energy efficiency is only about 55–66%. On the other hand, they are robust and have a long life, can be consistently overcharged, and suffer no damage if left in the discharged state. The electrolyte is liable to absorb $CO_2$ from the atmosphere, which reduces its conductivity, and it may need to be renewed occasionally.

*See also* Mi, P.

**Activation overpotential**

In any electrolytic process a part of the energy supplied is used in overcoming the internal resistance of the cell. If this is $R\,\Omega$ and the current passing is $I$ A, the corresponding voltage drop through the solution is $\Delta E = IR$ V. In addition, there will be a *concentration overpotential* (q.v.) at each electrode.

*Activation overpotential*

The remainder of the amount by which the applied e.m.f. $E$ exceeds the equilibrium e.m.f. $E°$ for the cell reaction is the activation overvoltage, which again is the sum of separate contributions from the two electrode processes:

$$E = E° + IR + \eta_{conc}(\text{cath}) + \eta_{conc}(\text{anode}) + \eta_{act}(\text{cath}) + \eta_{act}(\text{anode})$$

The two electrode potentials can be studied separately (see *decomposition voltage*) with elimination of the $IR$ term, so at a cathode

$$\Delta E = E° - E = \eta_{conc} + \eta_{act}$$

and under conditions where concentration overpotential is unimportant $\Delta E = \eta_{act}$ and similarly at the anode.

The values so measured vary very widely. For reactions which take place readily, such as the deposition of many metals, the cathode potential need be very little above its equilibrium value for large currents to pass; for other reactions, especially those in which gases are evolved, large overpotentials have to be established for useful currents to pass. It is presumed in such cases that there is a slow step in the discharge mechanism which causes a build-up of electrons at the cathode under steady state conditions (constant current).

*Hydrogen overpotential*

Hydrogen overpotential has been intensively studied; for hydrogen discharge from HCl solutions at a current density of $10^{-7}$ A m$^{-2}$ measured values vary from $\eta = 0.01$ V at platinised platinum to $\eta =$ 1.0 V at mercury, a 'high overpotential metal', showing that the electrode itself is critically involved in the process.

The general relation between current density and electrode potential was given by Tafel in 1905 (Tafel's equation):

$$j = \text{constant} \times \exp(-A\eta)$$

where A is a constant. Remembering that $j$ measures the rate of the reaction under the given conditions, the similarity to the Arrhenius equation, $k = \text{constant} \times \exp(-E_A/RT)$ becomes striking. Moreover, Tafel was able to show that, for hydrogen discharge at a mercury electrode, the factor A had a value equal to $F/2RT$, so that for this process the equation becomes

$$j = \text{constant} \times \exp(-F\eta/2RT)$$

The energy quantity $(F\eta/2)$ J here occupies the place of the activation energy $E_A$ in the theory of thermal chemical reaction rates and represents the energy per mole needed to surmount an energy barrier in the reaction path, i.e. to form the transition state. However, the factor $\frac{1}{2}$, which Tafel had to introduce, had no clear significance, and it has since been shown that in other circumstances the factor may have other values. The general form of the equation is therefore written

$$j = \text{constant} \times \exp(\alpha F\eta/RT) \tag{A.6}$$

where $\alpha$ is called the transfer coefficient.

Taking logarithms, equation (A.6) can be written

$$\log j = \log \text{constant} + 2.303(\alpha F\eta/RT) \tag{A.7}$$

or the experimental relation between $j$ and $\eta$ can be simply written as

$$\eta = a + b \log j \tag{A.8}$$

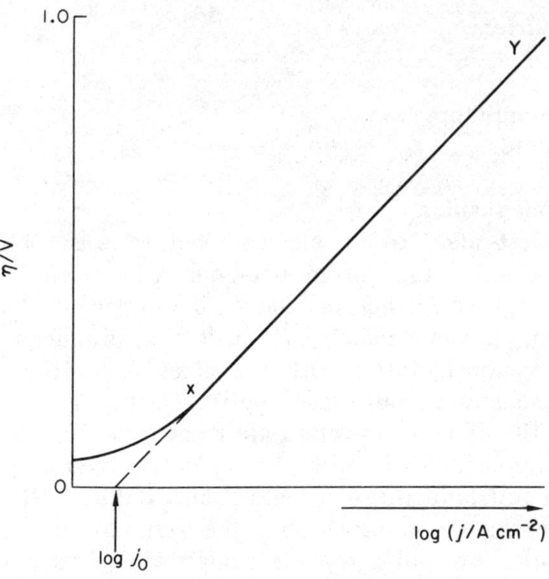


Figure A.2 Plot of the Tafel equation

where $a$ and $b$ are constants and $\eta$ is given a positive value; i.e. for a cathodic process, $\eta = E^\circ - E$.

A typical Tafel plot is shown in figure A.2. The Tafel relationship is obeyed between X and Y, and this straight line can be realised up to very high current densities if concentration overpotential is prevented. At very low current densities, however, the graph must always curve away to approach the limit $\eta = 0$ asymptotically for $j = 0, \log j = -\infty$. The slope of the line XY gives the value of $b$ in equation (A.8), and enables $\alpha$ to be evaluated ($b = 2.303\, RT/\alpha F$, from equation A.7). If the line is extrapolated back to $\eta = 0$, the value marked $\log j_0$ is obtained. The quantity $j_0$ is called the *exchange current density* (q.v.). This important constant provides further insight into the concept of activation overpotential, and into *electrode reaction mechanisms* (q.v.).

*See also* B & R, Fr.

**Activity**
*See* J.

**Activity coefficient**
*See* J.

**Alkaline accumulators**
*See* Accumulator.

**Alloy electrodeposition**
The commonest alloy to be electroplated is brass, which may be obtained in a wide range of copper–zinc ratios from cyanide baths. Surface coatings consisting of alloys of copper and cadmium, or containing tin, zinc and nickel, have also been produced.

The general considerations are the same as for the *electroplating* (q.v.) of single metals, but close control of the bath is necessary if variations in the alloy composition are to be avoided. The best conditions, and composition of the bath, are determined empirically. The discharge potentials of the two metals must naturally be very similar under the working conditions, but the standard potentials of the individual metals are not a reliable guide owing to the complexity of the solutions used and the effects on the overpotentials of the presence of the second metal in the cathode surface.

**Aluminium, electrometallurgy**
Aluminium is obtained on the large scale by electrolysing $Al_2O_3$, dissolved in a molten electrolyte, with graphite anodes. The over-all cell reaction is approximately

$$2Al_2O_3 + 3C \rightarrow 4Al + 3CO_2$$

the oxygen discharged at the anode combining with the carbon electrode; this reaction requires 12 F of electricity. The theoretical decomposition voltage for the reaction is only about 1.2 V, since much of the energy required to decompose the alumina is supplied by the oxidation of the carbon.

The choice of this system depends on a number of considerations. Aluminium has far too negative an electrode potential to be separable from aqueous solutions. The fused chloride is practically a non-conductor, and the low sublimation temperature of this salt is a further difficulty; other simple salts give anode reactions that make them unsuitable.

Aluminium oxide melts at 2000 °C, and the technical difficulties in working at this temperature would be almost insuperable. The oxide dissolves in cryolite, $Na_3AlF_6$, however, and the eutectic contains about 10% $Al_2O_3$ and melts below 1000 °C. Other salts can be added to reduce the m.p. still further, but they must not increase the density of the melt beyond 2.3, the density of molten aluminium, since the aluminium produced is collected as the lower layer to prevent oxidation. Magnesium fluoride and up to 5% calcium fluoride are common additions, and the working temperature is then not much above 900 °C.

It is not economic to prepare crude aluminium and then purify it, so very pure $Al_2O_3$ is used in the process, and the anodes should also be pure and ash-free. The product is aluminium of about 99.7% purity with traces of various impurities, mostly iron and silicon. (Even purer metal can be obtained from this by *electrorefining* (q.v.) in a bath similar to that used in the preparation. Aluminium alone ionises from the anode with 100% current efficiency.)

The cells consist of iron containers with carbon lining, which serve as cathode. Molten aluminium collects on the floor of the cell, where it can be tapped off. Above this is the fused electrolyte, into which the carbon anodes dip, with a crust of solidified electrolyte at its surface. The aluminium oxide consumed in the process has to be replaced periodically; when its concentration drops to a very low value, this is signalled by a rapid fall in the conductance. Finely powdered $Al_2O_3$ is then sprinkled on the surface, and dissolves rapidly when the surface is broken.

The exact composition of the electrolyte is unknown, and it could contain a large number of constituents: $AlO_2^-$, $AlF_6^{3-}$, and other oxygen and fluorine compounds of aluminium, as well as $Na^+$, $Al^{3+}$ and $F^-$. Consequently there is uncertainty about the primary cathodic and anodic reactions. The simplest cathodic reaction would be

$$Al_2O_3 + 3e \rightarrow Al + AlO_3^{3-}$$

At the anode the discharge of fluoride, or complex ions containing it, is very unlikely, so the probable reaction involves $AlO_2^-$ or $AlO_3^{3-}$, e.g.

$$2AlO_2^- \rightarrow Al_2O_3 + O + 2e, \quad \text{followed by} \quad 2O + C \rightarrow CO_2$$

The current efficiency of the process is about 80%, and the loss is due to the metal cloud that leaves the cathode. Sodium ions are present at very high concentration in the cathode double layer, and a little is probably discharged and leaves the cathode surface as minute bubbles of sodium vapour in which a little aluminium is entrained. The cloud diffuses or is carried electrophoretically to the neighbourhood of the anode, where it is oxidised by the carbon dioxide. This explanation is supported by analysis of the gases leaving the anode, which contain 30% or more of carbon monoxide.

*See also* Mi, P.

**Amalgam electrode**

The amalgam electrode is a variation of a metal–metal ion electrode in which the metal is in the form of an amalgam (i.e. dissolved in mercury) rather than in the pure form; a platinum wire is used to make electrical contact. The reaction occurring is the same as that at a metal electrode; the mercury plays no chemical role. In general, the potentials of amalgam electrodes are more reproducible than the potentials of solid metal electrodes, which are sensitive to surface impurities and mechanical strains in the crystalline solid. The contribution to the cell e.m.f. of an amalgam electrode is a function of the activity of the metal in the amalgam as well as of the metal ion in solution; they are represented diagrammatically:

$$X(\text{in Hg}, a_1) \mid X^+ \text{ aq } (a_2)$$

Amalgam electrodes are of value because they allow active metals, e.g. Na and K, to be used as electrodes. Reproducible, reversible potentials can be obtained at very low concentrations of the metal; it is best to use dropping electrodes, in which the amalgam drips slowly into the electrolyte, so that the surface is always fresh. The potential of the amalgam is compared with that of the pure metal by using both in a cell in which the electrolyte is dissolved in a solvent which does not attack the pure metal (e.g. ethylamine). Thus, to determine $E^{\ominus}(K^+, K)$, the sum of the standard e.m.f. values of the two cells:

$$\text{K, Hg} \mid \text{KCl} \mid \text{AgCl, Ag} \quad \text{and} \quad \text{K} \mid \text{KI in ethylamine} \mid \text{K, Hg}$$

gives the standard e.m.f. of the hypothetical cell:

$$\text{K} \mid \text{KCl} \mid \text{AgCl, Ag} \quad \text{as} \quad E^{\ominus} = 3.1464 \text{ V}$$

Hence, $$E^{\ominus}(K^+, K) = 3.1464 - 0.2224 = 2.9240 \text{ V}$$

*See also* Concentration cells; I & J.

**Ammonia probe**

The ammonia probe electrode has been designed for the measurement of ammonia concentrations in aqueous solutions and also the ammonium ion concentration once the sample has been converted to ammonia. The ammonia is detected by measuring the effect on the pH of an ammonium chloride solution, separated from the sample by a gas-permeable hydrophobic membrane (figure A.3). When the probe is in contact with the ammonia solution, the internal solution between the pH electrode membrane and the gas-permeable membrane gains or loses ammonia gas through the latter, until the partial pressure (activity) of ammonia is the same on both sides. Thus the pH of the internal solution is proportional to the free ammonia concentration in the sample.

The probe will be affected by other gases which have an acid or basic nature ($CO_2$, $H_2S$) in solution, when the gases have an effect on

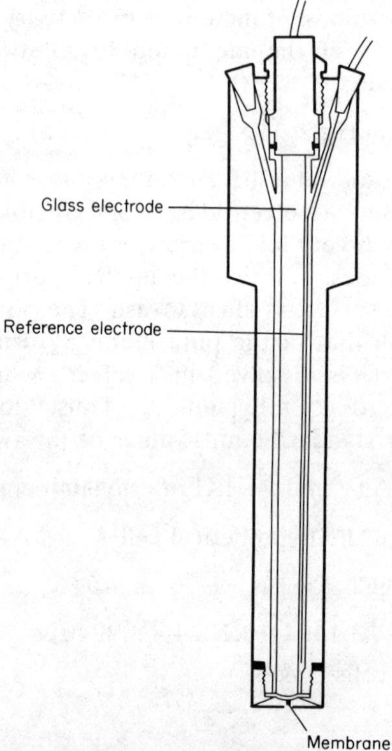


Figure A.3 Ammonia probe assembly

the internal filling solution. At high concentrations of ammonia, and, hence, at high pH values, these gases will not interfere. Since the gas-permeable membrane is hydrophobic, ions cannot enter the probe and will not therefore have any direct effect on the measurement. The concentration range is 0.1 to 1000 mg $dm^{-3}$. Although the probe is capable of detecting low levels of ammonia, it cannot, since it is a logarithmic device, detect very small changes in ammonia level at high concentrations. The potential of the probe is affected by change of temperature; hence, samples and standards should be at the same temperature (in the range 5–40 °C). Solutions containing known ammonia concentrations are used as standards and a calibration curve of potential against log concentration is constructed.

**Ampere**

The ampere, A, a basic SI unit, is that current which, if maintained in two parallel conductors of infinite length, of negligible cross-section, and placed 1 metre apart in vacuum, would produce between the conductors a force equal to $2 \times 10^{-7}$ newton per metre of length.

The international ampere is the current which deposits 1.118 00 mg of silver per second from a standard solution of silver nitrate.

The dimensions of current are $\varepsilon^{1/2} m^{1/2} l^{3/2} t^{-2}$.

*See also* Electric units.

**Amperometric titrations**

When a substance is undergoing reaction at an electrode in the presence of a large excess of an *indifferent electrolyte* (q.v.), it reaches the electrode as a result of diffusion, and at a sufficient overpotential the *limiting current density* (q.v.) for the reaction is reached. This is proportional to the bulk concentration of the electro-active substance, and can therefore be used as a measure of concentration. In an amperometric titration a suitable electrode, e.g. a rotating Pt electrode, is immersed in the solution under examination, and a constant voltage applied between it and a reference electrode (figure A.4). A suitable

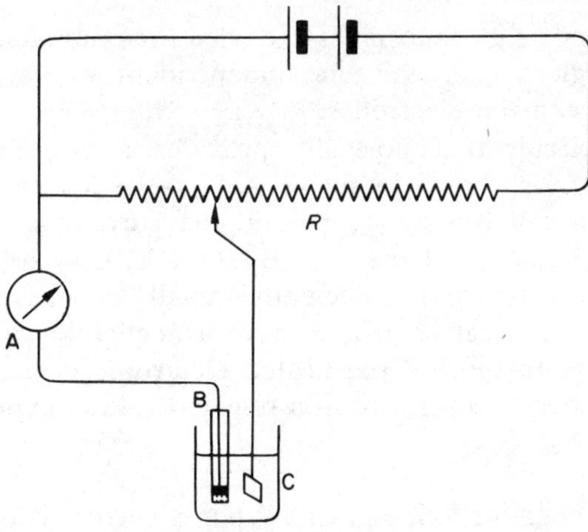


Figure A.4 Amperometric titration: R, rheostat; A, micro- or milliammeter; C, reaction vessel; B, reference electrode

reagent is then added, as in an ordinary volumetric titration, and the limiting current falls with each addition. At the end-point it becomes zero, and so remains after further additions, if the system and voltage are chosen so that no other electrode reaction can occur in the final solution. The titration diagram, current against quantity of reagent added, should consist, ideally, of two straight lines intersecting at the end-point.

The method has several points in common with a *conductimetric titration* (q.v.): (a) allowance must be made for the volume of reagent added if the branches of the diagram are to be straight lines; (b) the end-point is obtained by extrapolation of the branches, and can be measured very accurately; (c) slight deviations from linearity, which may occur around the end-point, need not influence the accuracy. Substances which do not undergo electrolytic reduction or oxidation (under the conditions of the experiment) can be titrated with a reagent which reacts at the electrode after the end-point has been reached. The second branch of the titration diagram in this case will be a line rising from the zero value of the first branch.

*See also* C, B & T, L.

**Amperostat**

The amperostat or galvanostat is a device for maintaining a constant current through a cell, and one independent of variations in the potential between the electrodes.

A simple circuit that normally provides a current constant to better than 1% is shown in figure A.5. A high resistance $R$ is in series with the cell (resistance $r$), and the current is provided by a high-tension battery (voltage $V$). By Ohm's law the current $I$ is $I=(V+\Delta E)/(R+r)$, and sufficiently small changes in $\Delta E$, the potential across the cell, and in $r$, have a negligible influence on the current passing. In more sophisticated electronic devices the current can be automatically controlled at a selected value to within 0.1%.

**Anion**

An anion is a negative ion, e.g. $Cl^-$. When a current is passed through an electrolyte solution, the anions move towards the positive electrode, the *anode* (q.v.).

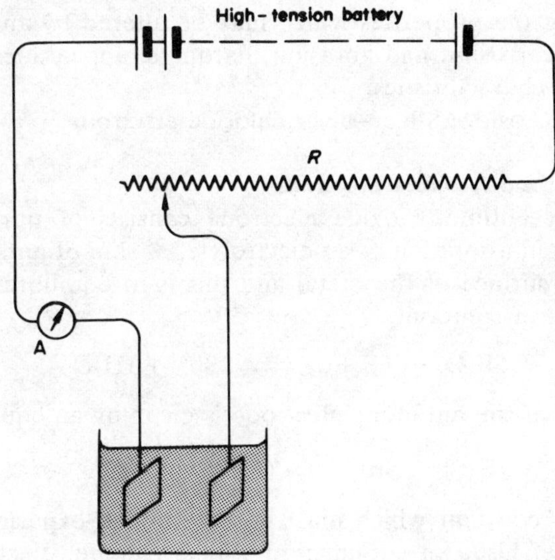


Figure A.5 Circuit for constant current measurements. A, ammeter

**Anode**

The anode is the positive terminal of a cell. Electrons tend to leave a cell through the anode, and to pass to the *cathode* (q.v.) through the external circuit. Within the electrolyte anions are attracted to the anode. It is the electrode at which oxidation reactions occur, e.g.

$$I^- \rightarrow \tfrac{1}{2}I_2 + e$$

$$Zn \rightarrow Zn^{2+} + 2e$$

$$Fe^{2+} \rightarrow Fe^{3+} + e$$

**Anodising**

Anodising is the process of forming oxide (and chloride) films or coatings on certain metals by electrolysis in a suitable solution. On the application of an electric potential to a cell in which the metal to be treated is the anode, the oxidising conditions convert the surface of the metal to the oxide. The oxide film is essentially an integral part of the metal. Among the properties which may be altered by anodising are: resistance to corrosion and abrasion, hardness, appearance, reflection and radiation characteristics.

*See also* Corrosion; Silver–silver chloride electrode.

**Antimony–antimony oxide electrode**

The antimony–antimony oxide electrode consists of pure antimony metal dipping in a solution of an electrolyte. A skin of antimony oxide forms on the surface of the metal and this is in equilibrium with the antimony ions in solution:

$$Sb_2O_3 + 6H^+ + 6e \rightleftharpoons 2Sb^{3+} + 3H_2O$$

The potential of the antimony electrode is given by an equation of the form

$$E(Sb^{3+}, Sb) = E' + (RT/F) \ln a(H^+)$$

where $E'$ is a constant which must be determined experimentally for each electrode. Used in conjunction with a calomel electrode and a potentiometer, this electrode measures the pH of a solution.

The electrode is very robust and can be used to determine the pH of solutions in the pH range 4–12 with an accuracy of ±0.2 pH unit; it does not contaminate the liquid under test, and since it has a low resistance, it can be used with a simple potentiometer. The electrode cannot, however, be used in the presence of dissolved oxygen, oxidising agents, hydrogen sulphide and heavy metal ions, or in highly acid or alkaline solutions. The *E*/pH conversion curve depends on the nature and concentration of the substances in the solution under test. The electrode is very susceptible to change of temperature.

*See also* Electrode; I & J, J & P.

**Arrhenius electrolytic dissociation theory**

Arrhenius (1882) was the first to advance the view that an electrolyte, when dissolved in water, dissociates extensively into free ions. His theory soon found support in the work of van't Hoff on the colligative properties of dilute solutions, since it was found that a salt such as NaCl had almost twice, while $CaCl_2$ had nearly three times, the effect on the vapour pressure, etc., of a solvent as a normal (undissociated) solute.

An electrolyte CA was thus supposed to exist in solution as an ionised fraction, $\alpha$, of free ions $C^+$ and $A^-$, in equilibrium with a fraction $(1-\alpha)$ of undissociated CA molecules. At infinite dilution dissociation would be complete (a necessary consequence of the law of mass action), and Arrhenius therefore proposed to determine $\alpha$ from the relation

$$\alpha = \Lambda/\Lambda^{\infty}$$

On this basis the dissociation constant of an electrolyte could be calculated from conductance measurements as follows:

$$K_{CA} = \frac{[C^+][A^-]}{[CA]} = \frac{\alpha^2 c^2}{(1-\alpha)c} = \frac{\Lambda^2 c}{\Lambda^{\infty}(\Lambda^{\infty}-\Lambda)}$$

This equation is often referred to as Ostwald's dilution law. It depends on the assumptions (a) that ionic conductivities have constant values, and are not dependent on the concentration; and (b) that the ions in dilute solution behave as ideal solutes. Both of these assumptions proved to be mistaken, and were finally corrected in the interionic attraction theory of Debye and Hückel and the conductance equation of Onsager (see *conductance of aqueous solutions*; *conductance equations*).

# B

**Battery**
*See* Accumulator.

**Beryllium, electrometallurgy**
Beryllium is obtained by the electrolysis of a fused mixture of sodium and beryllium chlorides. The anode is of graphite and the cathode is the nickel vessel in which the process is conducted. The operating temperature is about 670 K. At this temperature beryllium is solid and is removed from the cell walls from time to time.

**Bipolar electrode**

In some electrolytic processes (e.g. *electrolysis of water,* q.v.) metal plates unconnected with the electric supply are interposed between cathode and anode. Under the influence of the electric field, anions travel to these bipolar electrodes and are discharged, and the liberated electrons traverse the plate and initiate a cathodic reaction at the opposite surface. Compared with a series of separate cells, the system saves engineering and eliminates contact resistances.

**Bjerrum's ion-association theory**

Owing to the approximations made in its derivation, the *Debye–Hückel activity equation*‡ does not cover accurately the effects of interionic attraction between small ions at close distances of approach. To remedy this, Niels Bjerrum calculated the probability of finding an oppositely charged ion at given distances from a central ion. The probability is high for very small distances where electrostatic attraction is large, and passes through a flat minimum before increasing again as increasing volumes of solution are being considered. The minimum is at a distance $q$, given by $q = z_A z_B e^2/2kT$. For water at 298 K, $q = 3.5\, z_A z_B \times 10^{-10}$ m. Bjerrum proposed that a pair of ions at a distance of separation less than this should be treated as an uncharged *ion-pair* (q.v.) in equilibrium with the free ions (which, being separated by distances greater than the minimum, will obey the Debye–Hückel and Onsager equations accurately). The fraction of ion-pairs $(1-\alpha)$, can be calculated by integrating from $a$ (the distance of closest approach or mean ionic diameter) to $q$, and, hence, an association constant, the reciprocal of the dissociation constant,

$$K_D = \gamma_A \gamma_B \alpha^2 m/(1-\alpha)$$

can be obtained, where the ion acitivity coefficients, $\gamma_A$ and $\gamma_B$, are estimated by means of the Debye–Hückel equation, and the activity coefficient of the uncharged species is assumed to be equal to unity. Bjerrum's theory therefore brings short-range electrostatic forces within the framework of a dissociation–equilibrium formulation.

*See also* R & S.

**Brightening agents**
Brightening agents are small traces of organic compounds added to the salt solutions used in *electroplating* (q.v.) baths for electrowinning or *electrorefining* (q.v.) of metals. These compounds modify, mainly by adsorption processes, the crystalline growth of the deposited metal. The following are among the improvements that can be obtained: (a) to change coarsely crystalline deposits to microcrystalline deposits; (b) to decrease nodular growths; (c) to change the structure of the deposit; (d) to brighten the deposit; (e) to change the composition of alloy plates; and (f) to increase the hardness of the electrodeposit.

Thus the addition of glue to the acid bath used for the electrodeposition of zinc and cadmium decreases grain size and produces a smoother plate. In alkaline zinc cyanide baths glue is used in conjunction with other organic compounds to produce bright zinc plate which is used as a corrosion protection on ferrous articles. It is doubtless the great affinity of this large protein molecule for surfaces that affects the growing deposit.

Such agents are of importance in the plating baths used for the deposition of decorative as well as corrosion-protective coatings, e.g. bright nickel plate which confers protection to nickel–chromium coatings. The addition agents in this instance are more specific, more stable and more controllable in their effects than the colloidal type of agent; they are characterised by the presence of unsaturated bonds, e.g. dimethyl fumarate.

**Brønsted–Bjerrum equation**
*See* Ion-pair.

**Buffer solution**
*See* J.

# C

**Cadmium, electrometallurgy**
Cadmium is obtained by the electrolysis of cadmium sulphate solution, the process being very similar to that employed for zinc. Free acid is

formed during the electrolysis, and this is used, in a recycling process, to take more cadmium into solution. The main source of the cadmium is from the purification of zinc sulphate solution prior to its electrolysis; the cadmium is precipitated from this by the addition of zinc powder.

**Calcium, electrometallurgy**
Calcium is prepared by electrolysing fused calcium chloride at about 800 °C; some calcium fluoride may be added to lower the melting point. Chlorine is produced and is led off from the graphite anodes, and molten calcium is discharged at the metal cathode. It has a considerable solubility in the electrolyte, so a water-cooled cathode is used and the column of solidified calcium is continuously withdrawn from the electrolyte as electrolysis proceeds.

**Calomel electrode**
The calomel electrode is a reference electrode which consists essentially of mercury, mercury(I) chloride and potassium chloride solution of specified concentration (figure C.1), i.e. Hg, $Hg_2Cl_2$ | KCl (aq). The electrode potential is given by

$$E(\text{cal}) = E^{\ominus}(\text{cal}) + \frac{RT}{2F}\ln K_s - \frac{RT}{F}\ln a(\text{Cl}^-)$$

$$= E' - \frac{RT}{F}\ln a(\text{Cl}^-)$$

where $K_s = a(\text{Hg}_2^{2+})a^2(\text{Cl}^-)$.

The electrode thus behaves like a reversible chlorine electrode; the electrode potential, *E*, depends on the concentration of the potassium chloride solution and the temperature, *T*/*K*:

| | |
|---|---|
| KCl (0.1 mol $dm^{-3}$) | $E = 0.3335 - 0.000\,07(T - 298)$ |
| KCl (1.0 mol $dm^{-3}$) | $E = 0.2810 - 0.000\,24(T - 298)$ |
| KCl (saturated) | $E = 0.2420 - 0.000\,76(T - 298)$ |

The first electrode is preferred for accurate work as it has the lowest temperature coefficient; the saturated electrode is the most convenient owing to the ease of replacing the solution.

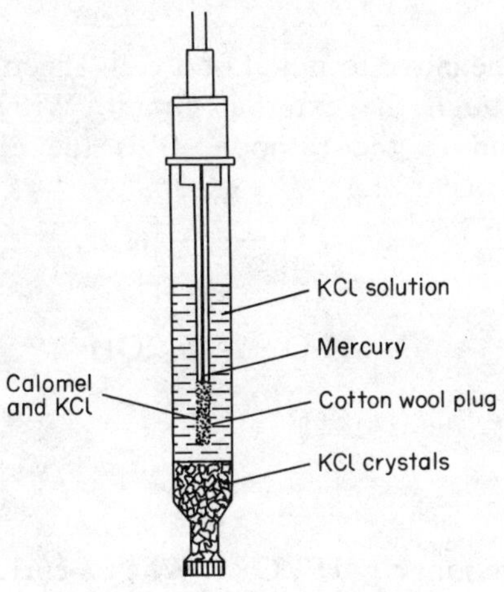


Figure C.1 Calomel electrode

Various types of calomel electrode are available commercially, while others can easily be prepared using a *salt bridge* (q.v.) to join the half-cell to the solution under test. In commercial electrodes the liquid junction is generally made by leakage of KCl solution through a ceramic disc.

*See also* I & J.

**Capacitance**

The capacitance of a conductor is defined as the ratio of its charge $Q$ to its potential $V$, i.e. $C = Q/V$. The unit is the farad, symbol F, measured in coulomb/volt. The capacitance of a parallel-plate capacitor is given by $C = (A\varepsilon_0\varepsilon/d)$ F, where $A$ is the area in $m^2$, $d$ the separation of the plates in m and $\varepsilon$ the dielectric constant of the medium between the plates.

In the CGS system $C = (A\varepsilon/36\pi d)$ F, where $A$ and $d$ are now measured in cm.

*See also* Conductance of aqueous solutions; Electrical double layer.

**Cathode**

The cathode is the negative terminal of a cell. Electrons tend to arrive at the cathode through the external circuit. Within the electrolyte cations are attracted to the cathode. It is the electrode at which reduction reactions occur, e.g.

$$H^+ + e \rightarrow \tfrac{1}{2}H_2$$

$$\tfrac{1}{2}I_2 + e \rightarrow I^-$$

$$\tfrac{1}{2}O_2 + H_2O + 2e \rightarrow 2OH^-$$

**Cathodic protection**

*See* Corrosion.

**Cation**

A cation is a positive ion, e.g. $H^+$, $Ca^{2+}$. When a current passes through an electrolyte solution, the cations move towards the negative electrode, the *cathode* (q.v.).

*See also* Anion.

**Cation-selective electrode**

*See* Ion-selective electrode.

**Cell**

All voltaic cells consist of a series of conducting phases in contact; the electrodes are generally metallic and there are one or more liquid electrolytes. At any phase boundary, where two or more phases of different composition meet, there is a difference of potential. The e.m.f. of the cell is the algebraic sum of all these phase-boundary potentials, including any metal contact potentials that may be present. The e.m.f. of a cell is the potential difference between two pieces of metal of identical composition, the ends of the chain of conducting phases.

Several types of cell are recognised:

(a) A primary cell is a device in which a spontaneous chemical reaction is used to produce electric energy. It acts as a source of electric energy until its materials are exhausted. Examples are the *Daniell cell* (q.v.), the *dry cell* (q.v.) and the *mercury cell* (q.v.).

(b) An *electrolytic cell* (q.v.) is the reverse arrangement, in which electric energy from an external source brings about the desired chemical reaction (see *electrodeposition of metals, electrolysis of water*).

(c) A secondary cell, or *accumulator* (q.v.), is a device which can act as a primary cell until it is exhausted (discharged), and can then be recharged and brought back to its original condition by passing through it electricity from an external source. It is acting as an electrolytic cell during the charging process.

Both primary and secondary cells can be connected in series to form a battery, which provides higher voltages than are obtainable from a single cell.

(d) A *fuel cell* (q.v.) is a primary cell in which the reagents can be continuously introduced, and the products of the reaction continuously removed, so that in theory the cell will function for indefinitely long periods.

(e) A *reversible galvanic cell* (q.v.) is specially designed for obtaining thermodynamic data, and for similar experimental uses. In any primary cell the whole free energy decrease of the reaction taking place can, ideally, be made available as electric energy, but in practice 'frictional' losses arise, both in the electrolyte (owing to its resistance) and at the electrodes (*overpotential*, q.v.). In reversible cells these losses are avoided by measuring potentiometrically the maximum e.m.f. to which the cell reaction could give rise under thermodynamically reversible conditions. The emphasis in their design is therefore not on their convenience as sources of electricity, but on the reproducibility and chemical purity of their constituent parts (see *thermodynamics of cells*).

(f) A *concentration cell* (q.v.), in which there is no over-all chemical reaction; the e.m.f. results from differences in concentration at the electrode surfaces of one or more of the reactants.

*See also* De, I & J, Mo.

**Cell constant**

The relation between the measured conductance of a solution and its conductivity (specific conductance) depends on the geometry of the cell employed in the measurement. It is impracticable to use cells for which this can be determined precisely, and cells are therefore calibrated with one of the accepted standard solutions (see *standards, conductance*).

*Cell constant*

The cell constant $C$ is defined by the relation

$$(1/R_{\text{expt}}) \times C = \kappa$$

The variation of the cell constant with temperature can be calculated from the coefficients of expansion of the cell material; over a few degrees it is usually quite negligible.

**Chloride cell**

In most primary cells the cathode reaction is the reduction of oxygen or an oxide, but in chloride cells a metal chloride is reduced to metal at the cathode. The special uses of the cells are for marine applications, as emergency power supplies, when the electrolyte can be sea-water. The anode is magnesium, the anode reaction being $Mg \rightarrow Mg^{2+} + 2e$. The cathode is separated from this by a porous diaphragm, and consists of a mixture of silver and silver chloride; the electrode reaction is $AgCl + e \rightarrow Ag + Cl^-$. The silver–silver chloride electrode can be replaced by copper–copper(I) chloride, which is cheaper but gives a lower e.m.f. The voltage of the silver chloride cell is about 1.7 V, falling to about 1.3 V at high discharge rates.

**Chlorine electrode**

*See* Gas electrode; Reversible galvanic cell.

**Chronoamperometry**

Chronoamperometry is the name given to techniques in which the potential at an electrode is controlled, and current–time curves are recorded. The solution is unstirred, and contains a swamping concentration of an *indifferent electrolyte* (q.v.), so that the electroactive constituent must reach the electrode by diffusion.

In linear chronoamperometry the potential is made to vary linearly with time, at a rate of 0.1–2 V $\text{min}^{-1}$. A curve such as that shown (figure C.2) is then obtained. The reaction begins when the decomposition potential is reached, and its rate, and therefore the current, rise rapidly with increasing overpotential. The concentration of the reactant at the electrode surface becomes depleted as the result of this, however, and can only be replaced by diffusion. The current cannot, therefore, increase indefinitely, but after a certain time reaches a

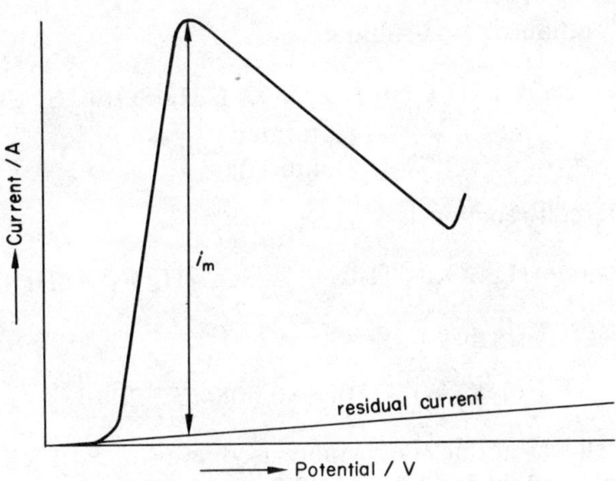


Figure C.2 Linear chronoamperometric curve

maximum and then decreases. With rising potential some further electrolytic process, such as decomposition of the solvent, eventually becomes possible, and the current will increase once more.

To a first approximation the maximum current, $I_m$, is given by

$$I_m = kAn^{3/2}D^{1/2}v^{1/2}c$$

where $k$ is a constant, $A$ the area of the electrode, $n$ the number of electrons transferred in the process, $D$ the diffusion coefficient, $v$ (= d$E$/d$t$) the rate of change in the potential and $c$ the concentration. The maximum current is therefore proportional to the concentration, and its height can be increased by using a faster rate of voltage increase. The method is an accurate one for estimating substances at low concentrations, and by a suitable series of experiments the constituents of a mixture can be estimated in turn.

*See also* C, B & T, L.

**Clark cell**

The Clark cell is a standard cell, in which mercury forms the positive pole and a 10% zinc amalgam the negative pole. The electrolyte is a saturated solution of zinc sulphate.

$$\ominus\ \text{Hg/Zn} \mid \text{ZnSO}_4.7\text{H}_2\text{O(s)} \mid \underset{\text{(saturated solution)}}{\text{ZnSO}_4} \mid \underset{\text{(paste)}}{\text{Hg}_2\text{SO}_4} \mid \text{Hg}\ \oplus$$

in which the cell reaction is

$$\text{Zn(s)} + \text{Hg}_2\text{SO}_4 + 7\text{H}_2\text{O} \rightarrow \text{ZnSO}_4.7\text{H}_2\text{O(s)} + 2\text{Hg(l)}$$

The e.m.f. at $T$ K is given by

$$E(T\ \text{K}) = 1.4330\{1 - 0.0084(T - 288)\}$$

The Clark cell was at one time widely used as a standard cell, but it has now been rejected, in favour of the *Weston cell* (q.v.), owing to its high temperature coefficient of e.m.f.

**Complex ions**
*See* Fused salts; Non-aqueous solutions; Transport number.

**Concentration cell**
A concentration cell is a cell in which there is no over-all chemical reaction, for the reaction occurring at one electrode (or pair of electrodes) is reversed at the other (or other pair of electrodes). There may, nevertheless, be a net change of free energy because of a difference in the concentration of one or other of the reactants concerned at the electrodes. The electrical energy arises from the *free energy*‡ change accompanying the transfer of material from one concentration to the other. The following general types of cell are recognised.

*Concentration cells with transport*
In these there is direct transfer of ions across liquid junctions.

(a) Simple cells containing a salt bridge:

$$\ominus\ \text{Ag} \mid \underset{m_2}{\text{AgNO}_3} \parallel \underset{m_1}{\text{AgNO}_3} \mid \text{Ag}\ \oplus \qquad m_1 > m_2$$

The double line in the centre represents the elimination of the *liquid junction potential* (q.v.); this can be achieved experimentally by interposing a *salt bridge* (q.v.):

$$\ominus \ \mathrm{Ag} \mid \underset{m_2}{\mathrm{AgNO_3}} \ \vdots\ \mathrm{NH_4NO_3} \ \vdots\ \underset{m_1}{\mathrm{AgNO_3}} \mid \mathrm{Ag} \oplus$$

At the right-hand electrode (RH), reduction occurs: $Ag^+ + e \rightarrow Ag(s)$. At the left-hand electrode (LH), oxidation occurs: $Ag(s) \rightarrow Ag^+ + e$. The net result is the transfer of $Ag^+$ from a solution of molality $m_1$ to one of molality $m_2$ and it is this that produces the e.m.f. The electrical neutrality of the solutions is maintained by the diffusion of anions or cations from the salt bridge. Since each electrode is a reversible silver electrode, it follows that

$$E(\text{cell}) = E(\text{RH}) - E(\text{LH}) = E^{\ominus}(\mathrm{Ag^+, Ag}) + \frac{RT}{F}\ln a(\mathrm{Ag^+})_1$$

$$- E^{\ominus}(\mathrm{Ag^+, Ag}) - \frac{RT}{F}\ln a(\mathrm{Ag^+})_2$$

$$= \frac{RT}{F}\ln\frac{a(\mathrm{Ag^+})_1}{a(\mathrm{Ag^+})_2} = \frac{RT}{F}\ln\frac{(a_\pm)_1}{(a_\pm)_2} \qquad \text{(C.1)}$$

assuming $a(\mathrm{NO_3^-}) = a(\mathrm{Ag^+})$.

As the reaction proceeds, the concentrations of the two solutions approach one another until they both have the same activity when $E(\text{cell}) = 0$.

Such cells are of use in the determination of the solubility and the *solubility product*‡ of sparingly soluble substances and in the determination of *activity*‡ values.

(b) Simple cells in which the liquid junction is not eliminated:

$$\ominus \ \mathrm{Ag} \mid \underset{m_2}{\mathrm{AgNO_3}} \ \vdots\ \underset{m_1}{\mathrm{AgNO_3}} \mid \mathrm{Ag} \ \oplus \qquad m_1 > m_2$$

For the passage of 1 F of electricity, transfer of material occurs as

*Concentration cell*

shown:

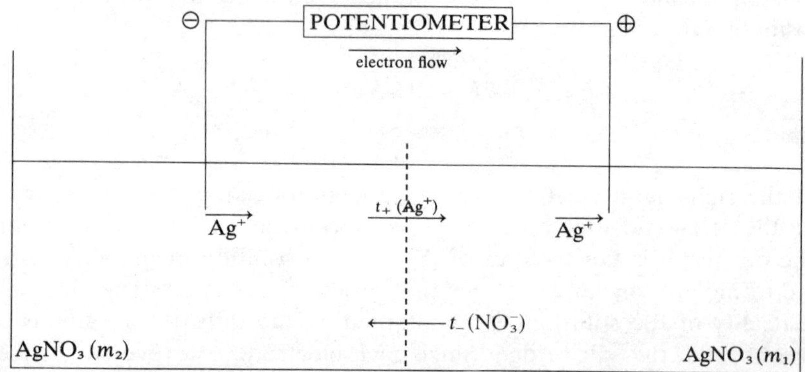


| | |
|---|---|
| At electrode: $Ag(s) \rightarrow Ag^+ + e$ | At electrode: $Ag^+ + e \rightarrow Ag(s)$ |
| At junction: $t_+$ mole of $Ag^+$ lost | At junction: $t_+$ mole of $Ag^+$ gained |
| Net gain: $(1-t_+)$ mole of $Ag^+ =$ $t_-$ mole of $Ag^+$ | Net loss: $(1-t_+)$ mole of $Ag^+ =$ $t_-$ mole of $Ag^+$ |
| Gain of $t_-$ mole of $NO_3^-$ by transfer | Loss of $t_-$ mole of $NO_3^-$ mole by transfer |

Thus over-all change is the transfer of $t_-$ mole of $Ag^+$ and $NO_3^-$ from $m_1$ to $m_2$. (It is assumed that the transport numbers, $t_+$ and $t_-$, of $Ag^+$ and $NO_3^-$, respectively, are constant in the concentration range $m_1$ to $m_2$.)

From a consideration of the *chemical potential*‡:

$$\Delta G = \sum_i n_i \mu_i$$

$$\Delta G = t_- RT \ln \frac{[a(Ag^+)a(NO_3^-)]_2}{[a(Ag^+)a(NO_3^-)]_1} = -FE$$

Thus $$E = 2t_- \frac{RT}{F} \ln \frac{(a_\pm)_1}{(a_\pm)_2} \qquad \text{(C.2)}$$

This is the complete e.m.f. including the junction potential. The *transport number* (q.v.) occurring in the equation is that of the ion to which the electrodes are *not* reversible. The general equation for the e.m.f. of a cell with electrodes reversible to positive ions is

$$E = t_- \frac{\nu}{\nu_+} \frac{RT}{nF} \ln \frac{(a_\pm)_1}{(a_\pm)_2}$$

The ratio of the e.m.f. of the cell in which the liquid junction is not eliminated (equation C.2) to that in which it is eliminated (equation C.1) gives approximate values for the transport numbers of the ions.

(c) Simple cells containing electrodes of the second kind:

$$\ominus \ \text{Ag, AgCl} \mid \underset{m_1}{\text{KCl}} \ \vdots\ NH_4NO_3 \ \vdots\ \underset{m_2}{\text{KCl}} \mid \text{AgCl, Ag} \ \oplus \qquad m_1 > m_2$$

At the RH electrode, reduction occurs:

$$AgCl(s) + e \rightarrow Ag(s) + Cl^-$$

At the LH electrode, oxidation occurs:

$$Ag(s) + Cl^- \rightarrow AgCl(s) + e$$

The net result is the transfer of $Cl^-$ from $m_1$ to $m_2$, for which

$$E = E(\text{RH}) - E(\text{LH}) = \frac{RT}{F} \ln \frac{a(Cl^-)_1}{a(Cl^-)_2} = \frac{RT}{F} \ln \frac{(a_\pm)_1}{(a_\pm)_2}$$

assuming $a(K^+) = a(Cl^-)$.

For the same cell in which the junction potential is not eliminated

$$\ominus \ \text{Ag, AgCl} \mid \underset{m_1}{\text{KCl}} \ \vdots\ \underset{m_2}{\text{KCl}} \mid \text{AgCl, Ag} \ \oplus \qquad m_1 > m_2$$

$$E = t_+ \frac{RT}{F} \ln \frac{(a_\pm)_1}{(a_\pm)_2}$$

*Concentration cells without transport*

These are cells in which liquid junctions are completely eliminated. The operation of these cells is not accompanied by direct transfer of electrolyte from one solution to the other, but occurs indirectly as a result of chemical reactions. Such cells result when two simple reversible galvanic cells whose electrodes are reversible with respect to each of the ions constituting the electrolyte are combined in opposition

*Concentration cell*

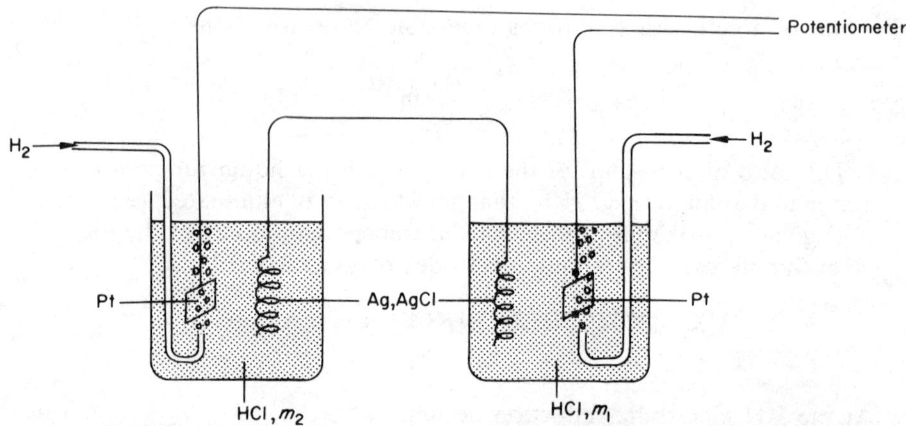


Figure C.3 Diagrammatic representation of a concentration cell

(figure C.3). Consider two cells of the type

$$\text{Pt, } H_2(g) \mid HCl \mid AgCl, Ag$$

(with different concentrations of HCl, $m_1$ and $m_2$, $m_1 > m_2$), each of which has the cell reaction

$$\tfrac{1}{2}H_2(g) + AgCl(s) \rightleftharpoons Ag(s) + H^+ + Cl^-$$

If they are now connected through the silver, in opposition,

$$\ominus\ \text{Pt, } H_2(g) \mid \underset{m_2}{HCl} \mid AgCl, Ag\text{—}Ag, AgCl \mid \underset{m_1}{HCl} \mid H_2(g), \text{Pt}\ \oplus$$

The over-all reaction, the sum of the two simple cell reactions, consists only of the transfer of HCl from $m_1$ to $m_2$ (assuming the hydrogen gas pressure is the same for both terminal electrodes):

$$\Delta G = [\mu(H^+)_2 - \mu(H^+)_1] + [\mu(Cl^-)_2 - \mu(Cl^-)_1]$$

$$= RT \ln \frac{a(H^+)_2 a(Cl^-)_2}{a(H^+)_1 a(Cl^-)_1} = 2RT \ln \frac{(a_\pm)_2}{(a_\pm)_1} = -FE$$

Hence, $$E = \frac{2RT}{F} \ln \frac{(a_\pm)_1}{(a_\pm)_2}$$

Such cells are used for the determination of activity values and for testing the validity of the *Debye–Hückel activity equation*‡. If the hydrogen gas pressure differs at the two electrodes, there will be an additional term in the above equation for the transport of hydrogen gas; see below.

*Electrode concentration cells*

(a) Gaseous electrodes of different pressures in the same solution:

$$\text{Pt, } H_2(\text{g}, p_1) \,|\, \text{HCl} \,|\, H_2(\text{g}, p_2)\text{, Pt}$$

The passage of 2 Faradays of electricity is accompanied by the transfer of 1 mole of hydrogen from $p_1$ to $p_2$; thus, assuming ideal behaviour,

$$E = \frac{RT}{2F} \ln \frac{p_1}{p_2}$$

If $p_2$ is kept constant while $p_1$ is varied, the equation takes the general form

$$E = \frac{RT}{2F} \ln p + \text{constant}$$

For hydrogen this equation is valid up to about 10 MN m$^{-2}$; deviations up to 60 MN m$^{-2}$ can be accounted for by making allowance for the deviations of hydrogen from ideal behaviour.

For the chlorine gas concentration cell,

$$E = -\frac{RT}{2F} \ln \frac{p_1}{p_2}$$

(b) Amalgam concentration cells:

$$\ominus \text{Zn/Hg}(x_1) \,|\, \text{ZnSO}_4 \,|\, \text{Zn/Hg}(x_2) \oplus$$

where $x_1$ and $x_2$ are the mole fractions of Zn in Hg, $x_1 > x_2$. At the RH electrode, reduction occurs:

$$Zn^{2+} + 2e \rightarrow \text{Zn}(x_2)$$

At the LH electrode, oxidation occurs:

$$\text{Zn}(x_1) \rightarrow Zn^{2+} + 2e$$

Since the concentration of $Zn^{2+}$ remains constant, the net change is the transfer of Zn from $x_1$ to $x_2$, for which

$$\Delta G = RT \ln a_2/a_1 \quad \text{and} \quad E = \frac{RT}{2F} \ln a_1/a_2$$

where $a_1$ and $a_2$ are the activities of zinc in the two amalgams. For the general case in which the valence of the metal is $z$ and there are $y$ atoms in the molecule,

$$E = \frac{RT}{zyF} \ln a_1/a_2 \approx \frac{RT}{zyF} \ln c_1/c_2$$

when the amalgams are sufficiently dilute. This type of cell is of use in determining the activity or concentration of metals in amalgams or alloys.

*See also* De, G, I & J, Mo.

**Concentration overpotential**

Concentration overpotential is one of the causes of overvoltage. It is present in all electrolytic processes, but can be studied without the complication of other effects by considering a cell in which two identical copper electrodes dip into a solution of a copper salt. In this symmetrical arrangement identical potentials are set up at the two electrodes, and if these are connected no current will flow; the system is in equilibrium. If now the electrodes are connected to an external source of e.m.f., however small, copper ions will be discharged at the cathode, and copper ions will pass into solution at the anode. The net chemical effect is nil, and for very small currents the energy supplied by the external source has only to overcome the resistance of the solution to the movement of the ions carrying current through the electrolyte.

At all but the lowest currents, however, concentration overpotential will become apparent. Of the copper deposited on the cathode, only a fraction $t_+$ arrives at the surface by electrolytic transport, and a fraction $(1-t_+)$ must be supplied, at first by the surface layer of solution and then by diffusion from more remote solution. From the beginning of electrolysis, therefore, there is a reduction of concentration round the cathode and a more negative potential will be needed to maintain the current; similarly, at the anode only a fraction $t_+$ of the copper ions dissolving are carried away by the current, the concentration will rise, and a more positive potential will be needed to cause more copper to dissolve.

The working cell has now become a *concentration cell* (q.v.) with concentrations $(c+x_1)$ and $(c-x_2)$:

direction of imposed (positive) current →

| Cu | $Cu^{2+}$ | $Cu^{2+}$ | $Cu^{2+}$ | Cu |
|---|---|---|---|---|
| anode | $c+x_1$ | $c$ | $c-x_2$ | cathode |
| | ← counter-e.m.f. | | | |

and if the electrolytic process were interrupted, a current would flow in the reverse direction under the influence of the 'counter-e.m.f.', and would continue until the concentration differences had disappeared. The applied e.m.f. has to overcome this counter-e.m.f., so that in the cell above there would be an overvoltage of

$$\Delta E = 2t_{-}(RT/F) \ln \{(c+x_1)/(c-x_2)\}$$

(assuming the cell resistance is unchanged, and neglecting activity coefficients).

In a working cell concentration overpotential will always be present, and may seriously reduce the efficiency of an electrolytic process. It can be greatly reduced by stirring the solution or by using rotating electrodes; even in an unstirred solution one does not have to rely entirely on diffusion to bring reactant ions to the electrode surface, as the density differences in the surface layers will lead to gravity mixing. A steady current will be maintained at a constant overvoltage if ions can be brought to the electrodes in these various ways as fast as they are used up by electrolytic action. There is a limit to this, however, set by the maximum rate at which ions can reach the electrode; the corresponding current cannot be exceeded (under the specified conditions) and this '*limiting current density*' (q.v.) is the maximum rate possible for the process.

## Conductance of aqueous solutions

### *Dilute salt solutions*

The high accuracy obtainable in conductance measurements can be maintained down to extremely dilute solutions; this is one reason for

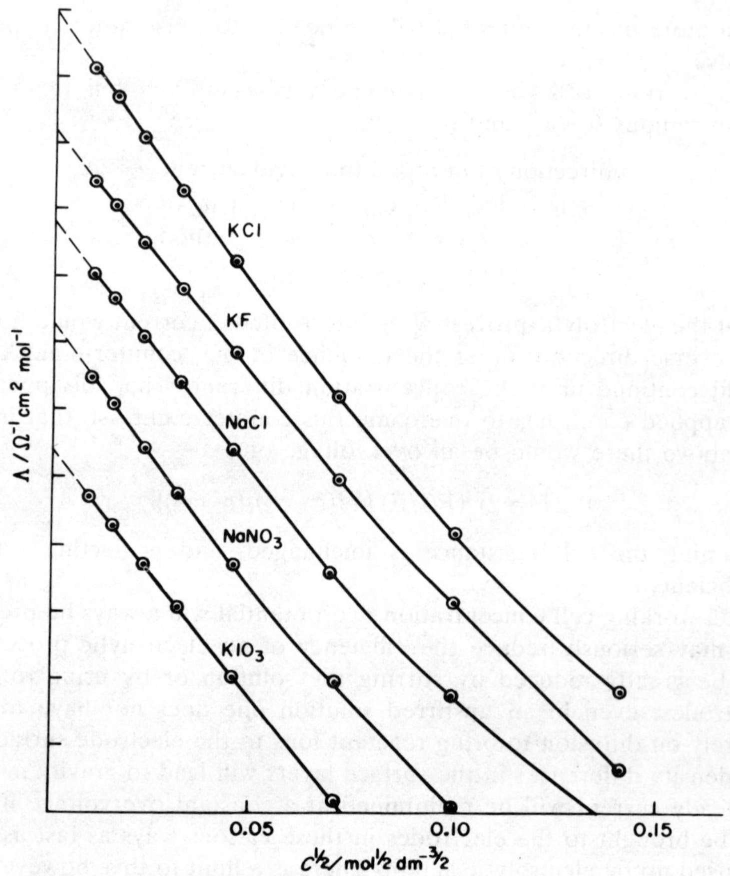


Figure C.4 Molar conductivities of uniunivalent salts in water at 291 K

the leading part conductance has played in the study of electrolytes. Figure C.4 shows some data for uniunivalent salts at 291 K; a similar figure could be constructed from the more recent results at 298 K. Up to a concentration of about 0.002 mol dm$^{-3}$ the $\Lambda$ values give straight lines when plotted against $c^{\frac{1}{2}}$, in agreement with the requirements of the Debye–Hückel–Onsager theory of interionic attraction. The lines only differ from one another in that the slope is higher the higher the $\Lambda$

value, and here, again, they are in quantitative agreement with Onsager's equation (see *conductance equations*). The agreement with the theoretical equation shows that these salts are completely dissociated into ions. At higher concentrations the lines curve away from the concentration axis, and individual differences begin to become more apparent, and in some cases there is evidence that ion-pairs tend to form as the concentration is increased.

When salts containing multivalent ions are considered in the same way, the slopes of the lines are steeper, owing to the stronger interionic forces, but the agreement with Onsager's equation is again very good for alkaline earth chlorides and several other bi-univalent salts, and even for the chlorides of lanthanum and the rare earth metals; but the experimental slopes for sodium and potassium sulphates are slightly steeper than the theoretical, and most salts containing ions of high valency show deviations from theory. Some examples are shown in figure C.5, including $CuSO_4$, which is typical of the bivalent metal sulphates. In extreme cases the curves are more typical of those of a weak electrolyte, and deviations of this kind are attributed to incomplete dissociation. This phenomenon is naturally more general among high-valency salts, and is reviewed in *ion-pair* (q.v.) and *Bjerrum's ion-association theory* (q.v.).

Conductance data are available for salts at much higher concentrations than those illustrated, but in the absence of a theory for concentrated ionic solutions little can be done in their interpretation.

*Weak electrolytes*

Conductance measurements provide a very sensitive way of studying the ionisation of a weak electrolyte. This applies more particularly to a binary compound such as an acid HX; for polyprotic acids such as $H_3PO_4$ the ionisation equilibria involve a number of ions and the interpretation of conductances is not straightforward.

As an example of the treatment, consider an acetic acid solution which at 298 K has a molar conductivity of 27.20 $\Omega^{-1}\,cm^2\,mol^{-1}$ at a concentration of $3.441 \times 10^{-3}$ mol $dm^{-3}$. The value at infinite dilution cannot be found by extrapolation, but data for the sodium salt obey Onsager's equation at high dilutions, and can be extrapolated to give $\Lambda^{\infty}(\text{NaAc}) = 91.00$. This can be combined with the known values for

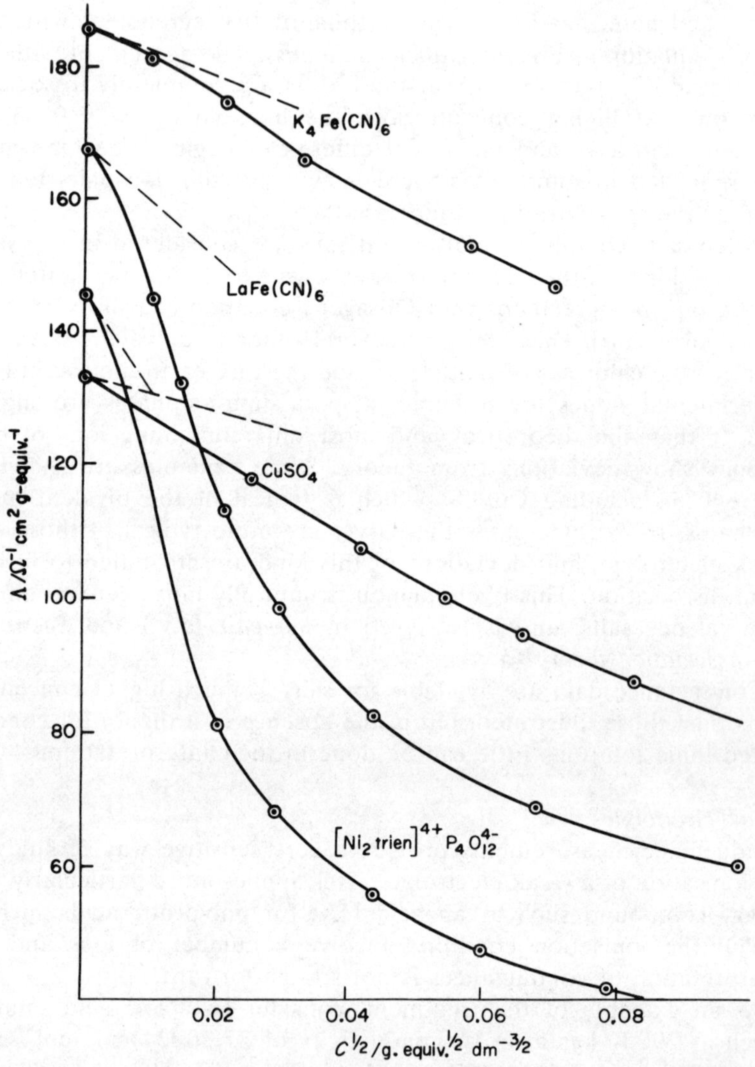


Figure C.5 Equivalent conductivities of some multivalent salts in water at 298 K. Dashed lines, Onsager slopes

$\Lambda^{\infty}(H^+) = 349.81$ and $\Lambda^{\infty}(Na^+) = 50.10$ to give

$$\Lambda^{\infty}(\mathrm{HAc}) = \Lambda^{\infty}(\mathrm{NaAc}) + \Lambda^{\infty}(\mathrm{H^+}) - \Lambda^{\infty}(\mathrm{Na^+}) = 390.71\ \Omega^{-1}\ \mathrm{cm^2\ mol^{-1}}$$

The quoted $\Lambda$ value is only about 7% of this, so even at $c = 0.003\ \mathrm{mol\ dm^{-3}}$ the ionisation of acetic acid is small. To obtain $\alpha$, the fraction ionised, allowance must be made for the fact that the conductrivities of the hydrogen and acetate ions depend on the ionic concentration, which is $\alpha c$. From Onsager's equation the sum of the ionic conductivities will be

$$\Lambda = 390.71 - 149.86(\alpha c)^{1/2}$$

When $c = 0.003\,441$, therefore

$$\alpha = 27.20/\{390.71 - 149.86(\alpha c)^{1/2}\} \qquad \text{(C.3)}$$

This is most simply solved by successive approximations. With $\alpha$ given the value 27.20/390.71, the right-hand side of equation (C.3) can be solved to give $\alpha = 27.20/388.39 = 0.070\,03$, and this value is unchanged by a second round of calculations. We can now calculate $K'$ in

$$K' = \frac{[\mathrm{H^+}][\mathrm{Ac^-}]}{[\mathrm{HAc}]} = \frac{\alpha^2 c}{(1-\alpha)} = \frac{(0.070\,03)^2 \times 3.441 \times 10^{-3}}{0.929\,97} = 1.815 \times 10^{-5} \qquad \text{(C.4)}$$

$K'$ is not the true dissociation constant, however, as the ions are not ideal solutes, and allowance must be made for their activity coefficients. These can be calculated from the *Debye–Hückel activity equation*‡, according to which,

$$\log \gamma(\mathrm{H^+})\gamma(\mathrm{Ac^-}) = -1.02(0.070\,03 \times 3.441 \times 10^{-3})^{1/2} \qquad \text{(C.5)}$$

and, combining equations (C.4) and (C.5), the correct dissociation constant is

$$\gamma(\mathrm{H^+})\gamma(\mathrm{Ac^-})\alpha^2 c/(1-\alpha) = 1.750 \times 10^{-5}\ \mathrm{mol\ dm^{-3}}$$

The older Arrhenius–Ostwald equation, which omitted both mobility and activity corrections, would have given the answer

$$K = \Lambda^2 c/\Lambda^{\infty}(\Lambda^{\infty} - \Lambda) = 1.79 \times 10^{-5}\ \mathrm{mol\ dm^{-3}}$$

and the weaker the electrolyte the more nearly correct will the

Arrhenius answer be. In some contexts, therefore, the labour of the extra calculations can be justifiably avoided (it may be noticed that this is because the two corrections partly cancel one another).

*Conductance of mixtures*

In a mixture of electrolytes all the ions present contribute to the ionic atmosphere of each ion, and it is the total concentration that governs the conductivity decrease. For salts that are not of the same valency type, the total concentration is replaced by the '*ionic strength*'‡, which is the quantity that determines the size of ionic atmosphere effects. This is given by $I = \frac{1}{2}\sum_i (z_i^2 c_i)$. That is, the ionic strength is half the sum of (charge number squared × concentration) of all ions present.

The conductance of mixtures always tends to be slightly less than the additive value given by the simple mixture rule. This is because the appearance in the ionic atmosphere of an ion of higher conductivity reduces the relaxation time effect, and vice versa. The effect is greater for the ion of higher conductivity, so the net result is a small conductance decrease.

If the electrolytes are not completely dissociated, mixing will alter the number of ions free to conduct, and there will be further effects that can be calculated if dissociation constants are known. This is illustrated (figure C.6) for mixtures in varying proportions of the reciprocal salt pair $NaCl$–$KNO_3$ at a concentration $c = 1\ \mathrm{mol\ dm^{-3}}$. Ion-pairing is appreciable in $KNO_3$ solutions, much less in $NaNO_3$, and absent in the chlorides. The main effect of adding $NaNO_3$ to $KCl$ is therefore to produce non-conducting $KNO_3$ ion-pairs, whereas diluting $KNO_3$ with $NaCl$ leads to some further dissociation of the nitrate. The continuous curves are calculated from the known dissociation constants. Much larger effects are possible with weaker electrolytes.

*Applications*

Conductance measurements have been used in studying the kinetics of reactions, including very fast reactions, and in a variety of other connections. They are particularly valuable in very dilute solutions, on account of their sensitivity, and in studying mixtures where the electrolyte is a very minor constituent. In analysis, a *conductimetric titration* (q.v.) can have some special advantages.

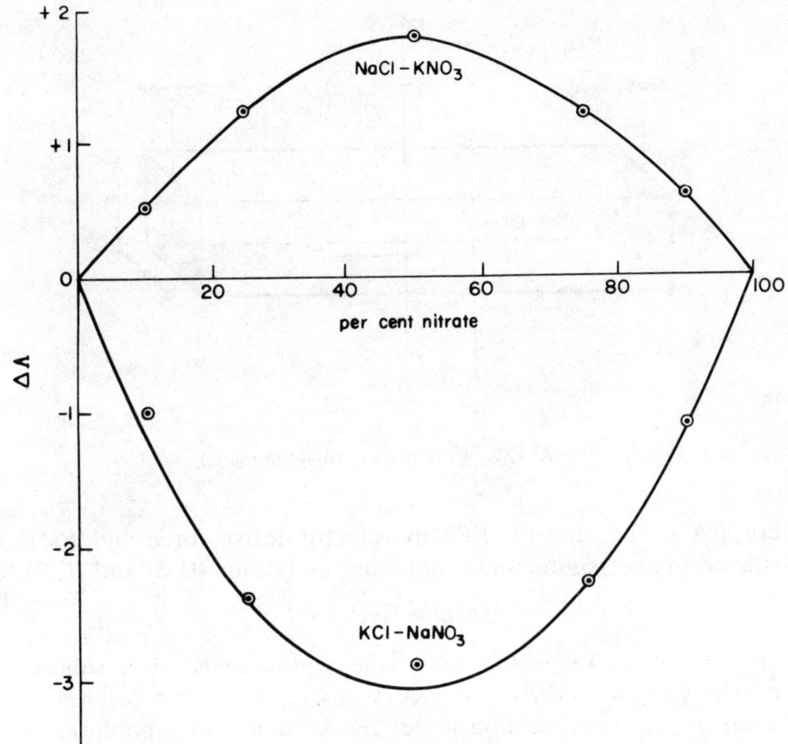


Figure C.6 Deviations from additivity in Na, K–Cl, $NO_3$ mixtures

*See also* D, H & O, R & S.

**Conductance, electric**

Conductance is the reciprocal of electric resistance. The resistance $R$ of a conductor is proportional to its length $l$, and inversely proportional to its cross-section $A$, and we can write this as

$$R = \rho(l/A) \tag{C.6}$$

where $\rho$ is the resistivity. The conductivity (formerly called specific conductance) is the reciprocal of this quantity: $\kappa = 1/\rho$.

Electrolyte solutions, like other conductors, obey Ohm's law:

$$I = E/R \tag{C.7}$$

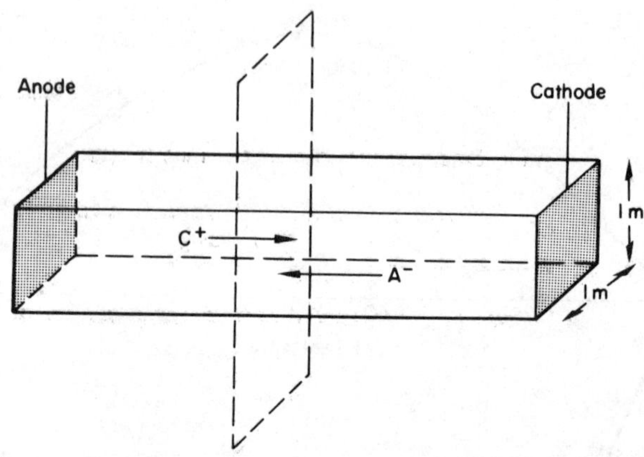


Figure C.7 Conductivity (diagrammatic)

where $I$/A is the current, $E$/V the electromotive force and $R/\Omega$ the resistance of the conductor. Combining equations (C.6) and (C.7),

$$EA/\rho l = E\kappa A/l = I \tag{C.8}$$

Putting $E$, A and $l$ equal to 1, the conductivity of a solution is numerically equal to the current in A passing through a column of the solution of unit cross-section under the influence of a potential gradient of 1 V per unit length.

Consider a column of electrolyte of cross-section 1 $m^2$ to which a potential gradient of 1 V $m^{-1}$ is applied (figure C.7). The current passing will depend on the number and speed of the ions and on the charges carried by them. All the ions present contribute and the total current will be the sum effect of the cations migrating towards the cathode and the complementary effect of negatively charged ions moving in the opposite direction. If a cation C carrying $z_C$ unit charges is present in the solution at a concentration $c_C$ mol $m^{-3}$, there will be $c_C$ mole in 1 m length of the column, and if their velocity under unit potential gradient is $u_C$ m $s^{-1}$, the quantity crossing some reference plane normal to the direction of the current in 1 s is $c_C u_C$ mole. One mole is associated with 96 490 $z_C$ C, so the contribution to the current is 96 490 $z_C c_C u_C$ A. By equation (C.8) the conductivity is given by the sum of all such contributions, and, hence,

$$\kappa = (96\,490\ z_C c_C u_C + 96\,490\ z_A c_A u_A + \ldots)\ \Omega^{-1}\ \mathrm{m}^{-1} \qquad \text{(C.9)}$$

The values of $u_C$, $u_A$, . . . are found to be around $5 \times 10^{-6}$ m s$^{-1}$. It is convenient to define the molar ionic conductivity by

$$\Lambda_i = 96\,490\ z_i u_i \qquad \text{(C.10)}$$

The general equation for the conductivity of a solution then becomes

$$\kappa = \sum c_i \Lambda_i\ \Omega^{-1}\ \mathrm{m}^{-1} \qquad \text{(C.11)}$$

where the summation is over all the ions present. This equation is a suitable starting point for most conductivity calculations. For a single electrolyte, CA, giving only one cationic and one anionic species, equation (C.11) becomes

$$\kappa = c_i(\Lambda_C + \Lambda_A) \quad \Omega^{-1}\ \mathrm{m}^{-1} \qquad \text{(C.12)}$$

The conductivity depends on the ionic concentration, and tends to zero as the solution becomes more dilute.

A more convenient unit for most purposes is the molar conductivity, obtained by multiplying the conductivity by the number of cubic metres containing 1 mole of the electrolyte, i.e. by dividing by $c$. From equation (C.12)

$$\Lambda = c_i(\Lambda_C + \Lambda_A)/c \quad \Omega^{-1}\ \mathrm{m}^2\ \mathrm{mol}^{-1} \qquad \text{(C.13)}$$

The molar conductivity, unlike the conductivity, is not directly dependent on the concentration, but it does depend on two factors that may vary with concentration changes. For an electrolyte that is not completely dissociated into ions it depends on $c_i/c$, the fraction dissociated. This is usually written as $\alpha$, which gives the general equation for a weak electrolyte:

$$\Lambda = \alpha(\Lambda_C - \Lambda_A) \quad \Omega^{-1}\ \mathrm{m}^2\ \mathrm{mol}^{-1} \qquad \text{(C.14)}$$

It also depends on the variation with concentration of the ionic conductivities; for a completely dissociated electrolyte this will be the only effect, and

$$\Lambda = (\Lambda_C + \Lambda_A) \quad \Omega^{-1}\,m^2\,mol^{-1}$$

Electrolytic dissociation becomes more complete as the concentration is reduced (this is a necessary consequence of the mass action law for any dissociation process). At the limit of infinite dilution, therefore, the molar conductivity will have a finite value given by

$$\Lambda^{\infty} = (\Lambda_C^{\infty} + \Lambda_A^{\infty}) \tag{C.15}$$

These limiting conductivities correspond to an ideal state in which the ions are so far separated as to have no mutual influence.

The relation between $\kappa$, $\Lambda$ and $\Lambda^{\infty}$ may become clearer by picturing a cell similar to that in figure C.7, with parallel electrodes 1 m apart, but in which the cross-sectional area of the electrodes, and the volume capacity of the cell, are infinitely extendable. If 1 mole of electrolyte dissolved in 1 $m^3$ solution is added to the cell, the current passing will measure $\Lambda$ (or $\kappa$, the two being the same for $c = 1$). If, now, successive additions of water are made to the cell contents, $\kappa$ (current carried per unit cross-section) will fall, and as the concentration approaches zero, so also will $\kappa$. There will still be 1 mole of electrolyte between the electrodes, however, and $\Lambda$ will actually rise, because in the more dilute solutions the ions have less of a retarding influence on one another, and also the electrolyte may be becoming increasingly ionised; at the limit of infinite dilution the conductivities will have their maximum values, shown in equation (C.15).

Most of the data in the literature are given in the CGS system of units, i.e. $\kappa$ (called specific conductance) will be in $\Omega^{-1}\,cm^{-1}$ units. To convert a specific conductance value into an SI conductivity value, multiply by $10^2$.

Moreover, the quantity of an ion usually considered was not the mole, but the gram-equivalent—that is, the quantity associated with 1 Faraday of electricity. Concentrations were therefore given in g-equiv. $dm^{-3}$, and equivalent conductivities, rather than molar conductivities, were quoted. To convert an equivalent conductivity to a molar conductivity, divide by $10^4$ and multiply by the number of electric charges associated with the ionised molecule (e.g. 6 for $Al_2(SO_4)_3$).

The two systems are illustrated by the following figures for magnesium chloride (at infinite dilution) in water at 298.15 K ($\Lambda$, the symbol for equivalent conductivity, has now been adopted as the symbol for molar conductivity.)

| Equivalent conductivity/ $\Omega^{-1}$ cm$^2$ g-equiv.$^{-1}$ | | Molar conductivity/ $\Omega^{-1}$ m$^2$ mol$^{-1}$ | |
|---|---|---|---|
| $\Lambda(MgCl_2)$ | 129.3 | $\Lambda(MgCl_2)$ | $258.6\times10^{-4}$ |
| $\Lambda(Mg^{2+})$ | 53.0 | $\Lambda(Mg^{2+})$ | $106.0\times10^{-4}$ |
| $\Lambda(Cl^-)$ | 76.3 | $\Lambda(Cl^-)$ | $76.3\times10^{-4}$ |
| | | $\Lambda(\frac{1}{2}MgCl_2)$ | $129.3\times10^{-4}$ |

*Measurement*

The resistance of the solution is measured in a suitable conductance cell, alternating current being used to avoid electrode reactions; the conversion into a true conductivity is effected by calibrating the cell with a standard solution (see *standards, conductance*).

A Wheatstone bridge is normally used (figure C.8). The source of alternating current S is an oscillator generating sine-wave a.c. at a range of frequencies,, say 500–5000 Hz. D is a current detector, which may be a telephone ear-piece, a rectifier and microammeter, or a cathode ray oscilloscope. Two arms of the bridge contain C, the cell, and a calibrated resistance box $R$; the other two arms are also wire

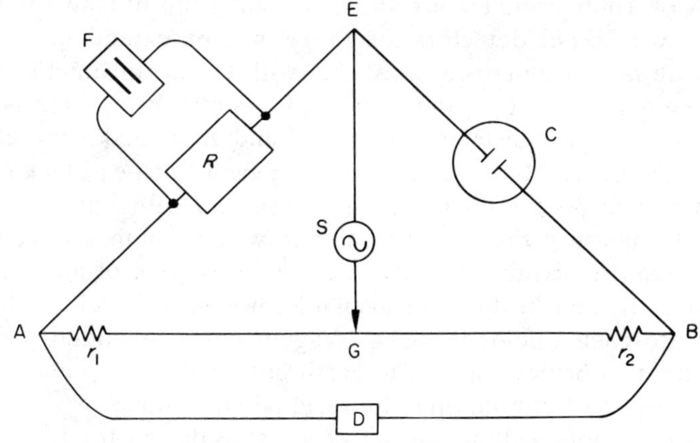


Figure C.8 Conductance bridge

resistances $r_1$ and $r_2$, G being a movable contact. If these components all behaved as pure resistances, the bridge could be balanced by adjusting the resistances until zero current passed through the detector. At this null-point A and B would be at the same instantaneous potential, and as the fall in potential through $R$ and $r_1$ is the same as through C and $r_2$ it follows that

$$\frac{\text{resistance } R}{\text{resistance } r_1} = \frac{\text{resistance of C}}{\text{resistance } r_2}, \quad \text{or resistance of} \quad \text{C} = \frac{Rr_2}{r_1}$$

The balance of an a.c. bridge is complicated, however, by inductance and capacitance effects. The former can be minimised by correct spacing and non-inductive winding in the resistance coils. The cell always possesses capacity (a) due to an electrical double layer of ions at the electrode surfaces; (b) between the cell and the surrounding thermostat; this can be minimised by using a light oil instead of water in the thermostat; (c) between the solution in the cell and the leads to the electrodes; this can be avoided by a wider spacing of the leads. To balance the remaining effects, a variable capacitor F is placed in parallel with $R$. A simultaneous setting of G and F should now give a very sharp end-point. A further criterion of a correct resistance balance is to repeat the measurement at a changed frequency, since the interfering effects are frequency-dependent. Using a telephone, frequencies of 1000–2500 Hz are those to which the human ear is most sensitive; with other detectors the range may be extended.

Accurate measurements are possible with the arrangement of figure C.8, using a telephone as detector, and a metre bridge for AB with extensions of high-resistance wire at A and B to make the effective length of the bridge 10 m or more; G may be a Perspex block carrying a fine wire and resting on the bridge wire, and with this it should be possible to measure the balance point to within 1 mm. Concentration and temperature errors are likely to exceed this error of measurement. More sophisticated bridges are available, however, in which the components are each shielded, and a 'Wagner earth' is added; this is an arrangement to bring A and B to earth potential at the balance point, with the object of eliminating any residual stray noise.

The conductance cell should be of suitable design for the range of resistances to be measured, which should preferably be in the range

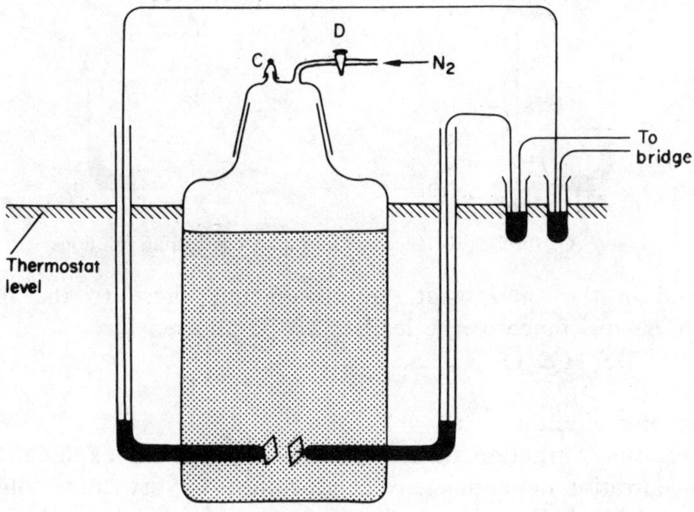


Figure C.9 Conductance cell for dilute solutions

500–5000 $\Omega$. For very dilute solutions, the modified Hartley–Barrett cell shown in figure C.9 is convenient. In this the conductance of the solvent is first measured, so that a *solvent correction* (q.v.) can be applied. The solution of solute is then added from a weight-burette through C while a pressure of purified air or $N_2$ is maintained through D. The height of the liquid in the cell must be sufficiently far above the electrodes for the measured resistance to be independent of it, and this is indicated by an etch-mark on the wall. The electrodes are usually platinised to reduce polarisation errors.

For more highly conducting solutions, cells such as that shown in figure C.10 are suitable. For conductimetric titrations dipping electrodes may be used.

The conductance of electrolytes increases with temperature by more than 2% per degree, so close temperature control of the thermostat is necessary. Also, generation of heat inside the cell must be minimised by making measurements quickly with very small currents. In addition, heat exchange between the electrodes and the bridge must be avoided, and this can be done by taking leads from the cell to tubes of mercury

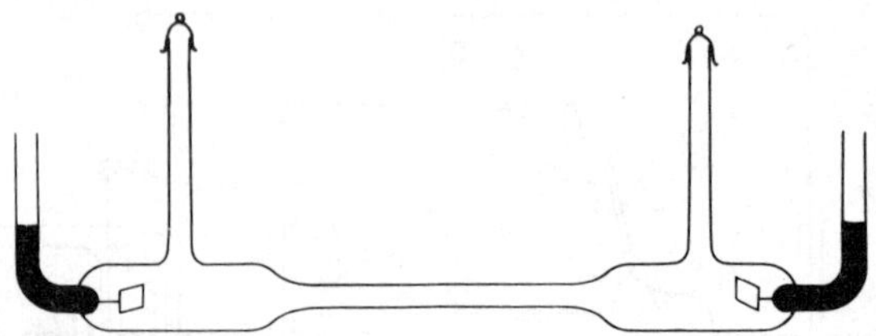

Figure C.10 Conductance cell for highly conducting solutions

immersed in the thermostat, and connecting these to the outside circuit. The resistance of the leads must be allowed for.

*See also* De, H & O, R & S.

**Conductance equations**

The interionic attraction theory provides a complete explanation for the concentration dependence of conductances in very dilute solutions. Debye and Hückel's original treatment was improved by Onsager in 1926 and his equation has been convincingly tested over a wide range of conditions.

Consider a dilute solution of a completely dissociated electrolyte, such as sodium chloride. If there were no forces between the ions, their thermal motions would keep them randomly and, on average, uniformly distributed through the solution. But according to Coulomb's law, there are forces of attraction or repulsion between them of $e^2/\varepsilon r^2$, where $e$ is the electric charge on a univalent ion, $r$ is the distance between them and $\varepsilon$ is the dielectric constant; these forces are appreciable over quite large distances. A given sodium ion will therefore attract the surrounding chloride ions and repel other sodium ions, so that at any point close to the ion there will be a greater likelihood of finding a chloride ion than a sodium ion. These directive forces are superimposed on the random thermal motions, but a time-average distribution can be calculated. Accordingly, each ion has in its close neighbourhood a slight excess of ions of the opposite sign. Each ion is thus said to possess an 'ionic atmosphere' of opposite sign, the effects of which will become less important as the solution is diluted, and will disappear at infinite dilution.

If the charge on an ion were suddenly annulled, its ionic atmosphere would no longer be attracted and would disperse. This, or the opposite process of building up an atmosphere, would take a finite time, depending on the diffusion rates of the ions involved. This is the relaxation time, and for these dilute solutions is about $10^{-6}$ s.

Its ionic atmosphere affects an ion's mobility in two ways. First, when an ion moves through the liquid, its ionic atmosphere will build up around its new position and die away around the position it has just left. Since this takes a finite time, the ion will always be moving away from a region containing excess of electricity of the opposite sign and so will be subjected to a retarding force. This is the 'relaxation time effect'. The second retarding influence is the 'electrophoretic effect'. The ions in moving through the solution transfer momentum to the solvent molecules; but each ion on account of its atmosphere is surrounded by an excess of those moving in the opposite direction, so that it will be moving 'up-stream', and will be opposed by a greater frictional force than stationary solvent would exert. Both retarding effects depend on the density of the ionic atmosphere, and increase as the square root of the concentration.

Onsager's equation for the equivalent conductivity of an ion in very dilute solution is

$$\Lambda_i = \Lambda_i^{\infty} - \left(\frac{8\pi N_A e^2}{1000\varepsilon kT}\right)^{1/2}\left[\left(\frac{z_1 z_2 e^2}{3\varepsilon kT}\right)\left(\frac{\Lambda_i^{\infty} q}{1+q^{1/2}}\right) + \frac{F^2 z_i}{6\pi\eta N_A}\right]\left(\frac{z_1 z_2}{2}\right)^{1/2} c_E^{1/2}$$

where

$$q = \frac{z_1 z_2(\Lambda_1^{\infty} + \Lambda_2^{\infty})}{(z_1 + z_2)(z_2\Lambda_1^{\infty} + z_1\Lambda_2^{\infty})}$$

Here $c_E$ is the equivalent concentration of an electrolyte, $z_1$ and $z_2$ the charge numbers of the two constituent ions, and $\Lambda_1^{\infty}$ and $\Lambda_2^{\infty}$ their equivalent conductivities; $\varepsilon$ is the dielectric constant of the solvent and $\eta$ its viscosity in poise. When the universal constants $e$, $N_A$, $F$ and $k$ are given their values, the equation becomes

$$\Lambda_i = \Lambda_i^{\infty} - \left\{\frac{2.801 \times 10^6 z_1 z_2 q \Lambda_i^{\infty}}{(\varepsilon T)^{3/2}(1+q^{1/2})} + \frac{41.25 z_i}{\eta(\varepsilon T)^{1/2}}\right\}\left(\frac{z_1 + z_2}{2}\right)^{1/2} c_E^{1/2} \quad \text{(C.16)}$$

For a completely dissociated electrolyte the equivalent conductivity is the sum of two such terms, and is thus

$$\Lambda = \Lambda^{\infty} - \left\{ \frac{2.801 \times 10^{6} z_1 z_2 q \Lambda^{\infty}}{(\varepsilon T)^{3/2}(1+q^{1/2})} + \frac{41.25(z_1+z_2)}{\eta(\varepsilon T)^{1/2}} \right\} \left( \frac{z_1+z_2}{2} \right)^{1/2} c_{\mathrm{E}}^{1/2} \quad \text{(C.17)}$$

For a binary electrolyte (one that dissociates into two ions) $q = \frac{1}{2}$, and the equation can be written

$$\Lambda = \Lambda^{\infty} - (a\Lambda^{\infty} + b)c_{\mathrm{E}}^{1/2} = \Lambda^{\infty} - Sc_{\mathrm{E}}^{1/2} \quad \text{(C.18)}$$

where $S$ is often called the 'Onsager slope', and $a$ and $b$ are constants with values given by equation (C.17); $a\Lambda^{\infty}$ gives the relaxation-time effect and $b$ the electrophoretic effect. For uniunivalent electrolytes in water at 298 K, $a = 0.2292$ and $b = 60.32$.

The equation reproduces the limiting slopes of $\Lambda$–$c_{\mathrm{E}}^{1/2}$ curves very satisfactorily over a wide range of conditions. For instance, for KCl the calculated slopes agree with the experimental at 273 K ($S = 47.3$) and at 373 K ($S = 313.4$); the equation also reproduces the large effects of changing valency (see *conductance of aqueous solutions*) and of variations in the dielectric constant and viscosity of the solvent (see *non-aqueous solutions*). The equation is accurate at concentrations up to $c_{\mathrm{E}} = 0.002$ g-equiv. dm$^{-3}$ for uniunivalent salts in water. For multivalent salts and for electrolytes in non-aqueous solvents, its range is more restricted.

*Extensions to higher concentrations*

The extension of the theory to higher concentrations has been undertaken particularly by Falkenhagen and his co-workers, by Pitts, and by Onsager and Fuoss. The theory is very difficult, and the final equations inevitably involve assumptions and mathematical approximations that have been the subject of discussion; they are too complicated to reproduce here, and require computer time for their solution. All three can satisfy the best experimental results for uniunivalent salts up to 0.1 g-equiv. dm$^{-3}$; for unsymmetrical valence types the mathematical problems are intractable, and above 0.1 g-equiv. dm$^{-3}$ for any salt the model on which they are based becomes unreliable.

The main requirement in extending the theory is the introduction of a parameter $a$, the mean distance of closest approach, or mean ionic diameter, to allow for the fact that ions have finite size and are not the point charges assumed in deriving the limiting laws. The value of $a$ is chosen to give the best fit with the data, the only requirement being that it should be a plausible value ($\sim 10^{-10}$ m) for the effective diameter of a dissolved ion. The different theories show variations in their precise definition of $a$, and require different values to satisfy the same data.

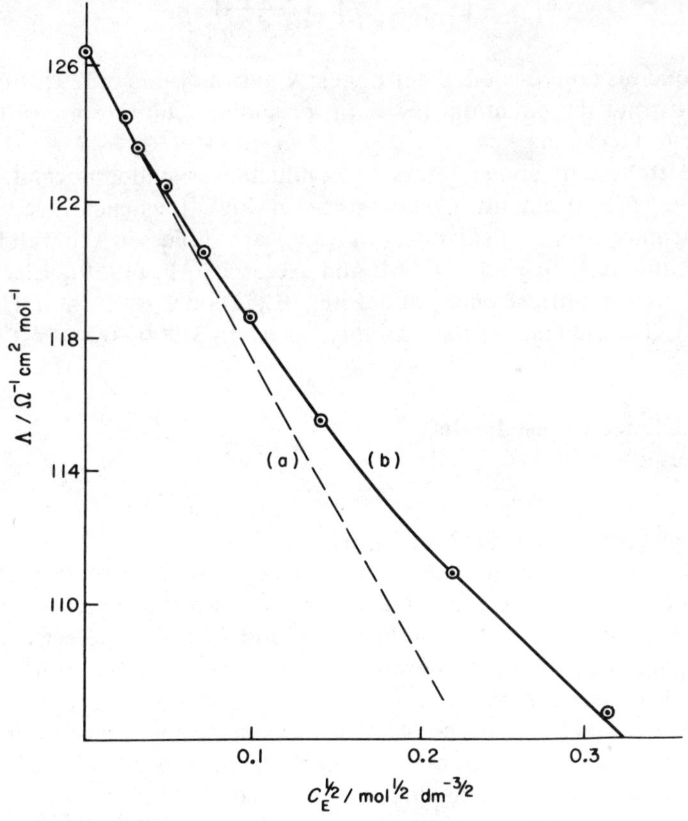


Figure C.11 NaCl at 298 K. Curve (a), Onsager's equation; curve (b), Robinson and Stokes's equation ($a = 4 \times 10^{-10}$ m)

*Conductance equations*

For practical purposes, the most useful equation is that of Robinson and Stokes, who have shown that the Falkenhagen equation can be approximated to

$$\Lambda = \Lambda^{\infty} - Sc^{1/2}/(1+\kappa a) \qquad \text{(C.19)}$$

which is simply the limiting equation with the Onsager term divided by $(1+\kappa a)$, where

$$\kappa = \left(\frac{8\pi N_A e^2}{1000\varepsilon kT}\right)^{1/2}\left(\frac{z_1+z_2}{2}\right)^{1/2} c_E^{1/2}$$

For aqueous uniunivalent electrolytes, $\kappa$ has the value $0.3291\times10^8\ c_E^{1/2}$. To illustrate the equation, the data for sodium chloride are plotted in figure C.11.

The treatment above refers to conductances under normal conditions of measurement. *Conductance at high frequencies* (q.v.) and *conductance at high field strengths* (q.v.) are discussed separately.

*See also* R & S; Fuoss, R. M. and Accascina, F. (1959), *Electrolytic Conductance*, Interscience; and Pitts, E., Tabor, B. F. and Daly, J. (1969), *Transactions of the Faraday Society*, **65,** 849–62.

**Conductance of fused salts**
*See* Fused salts

**Conductance at high field strengths**
In 1927 Wien found that the conductance of an electrolyte changes if very high voltages are applied. With strong electrolytes there is a moderate increase in the conductance, and with weak electrolytes an additional effect that may be much larger. These are sometimes called the first and second Wien effects.

Under normal conditions of measurement an ion migrates at a rate of, perhaps, 2 cm $h^{-1}$, whereas at a field strength of 200 kV $cm^{-1}$ it will move at a speed of some metres per second. Under these extreme conditions the ionic atmosphere of the ion cannot re-form, and the relaxation-time effect disappears; the conductance is therefore greater than that given by the Onsager equation (see *conductance equations*).

A quantitative theory of the Wien effects has been given by Onsager and Wilson; they show that a small electrophoretic effect remains, so that at high voltages the molar conductivity of a dilute solution increases to a value only a little below the normal $\Lambda^{\infty}$ value.

The second Wien effect is easily understood by remembering that for the normal equilibrium of a weak electrolyte, $CA \rightleftharpoons C^{+} + A^{-}$, the ions are surrounded by their ionic atmospheres, comprising an excess concentration of the ion of opposite sign. When this excess concentration is dispersed, the equilibrium will be displaced further to the right, so that at high field strengths the fraction ionised is greater. Incidentally, a further effect of the absence of ionic atmospheres is that in calculating the enhanced dissociation constant of the weak electrolyte, the activity coefficients of the ions may be put equal to unity, thus removing one of the uncertainties normally present in p$K$ measurements.

To avoid a large temperature rise in the solution, the passage of current at high voltages must be extremely brief. Wien used a highly damped condenser discharge, but Patterson has since adapted modern techniques of single-pulse generation and measurement to obtain results of high accuracy, which are in good agreement with the theory. As an example, the fraction of free ions in $ZnSO_4$ of concentration $1.64 \times 10^{-4}$ mol dm$^{-3}$ increases from a 'normal' value of 0.969 to 0.994 at 200 kV cm$^{-1}$.

*See also* F, H & O.

**Conductance at high frequencies**

Electrolytes show an enhanced conductance when the measurements are made at very high frequencies. This is because the relaxation-time term of the Debye–Onsager equation (see *conductance equations*) must diminish, and eventually disappear, at very high frequencies. The time taken to establish the ionic atmosphere around an ion is about $10^{-6}$ s in dilute solutions, and if the frequency used is so high that the ion is oscillating at a similar or greater rate, the atmosphere about the ion will not be fully established, and its retarding effect will be reduced.

The quantitative treatment of the effect has been worked out by Falkenhagen, and this has been confirmed by experiment; at wavelengths of a few centimetres the relaxation-time effect is almost completely absent. As an example, some results for KCl solutions at 298 K are given in the table:

Conductance values/$\Omega^{-1}$ cm$^2$ mol$^{-1}$ at high frequencies

| $10^4$ c/mol dm$^{-3}$ | $\Lambda(f=0)$ | $\Lambda(f=37.5$ MHz$)$ |
|---|---|---|
| 4 | 148.14 | 148.62 |
| 8 | 147.39 | 147.95 |
| 16 | 146.39 | 147.00 |

*See also* F, H & O.

**Conductance minima**

Occasionally the curve of equivalent conductivity against concentration is found to pass through a minimum value. This behaviour is common for salts in pyridine, sulphur dioxide and the aliphatic amines, all solvents with dielectric constants around 10. It also occurs in aqueous solutions of lanthanum ferricyanide and of the colloidal electrolytes.

There is no reason for thinking that ionic conductivities could increase with concentration, so the explanation must lie in a change in the number or nature of the ions present. Actually, it is unnecessary to postulate any radical change in the nature of the ionisation process. The dissociation of a partly dissociated binary electrolyte is governed by the equation

$$\frac{\gamma_+\gamma_-\alpha^2 c}{1-\alpha}=K \qquad \text{(C.20)}$$

If, over a concentration range, $\gamma_+\gamma_-$ decreased more rapidly than the increase in $c$, the value of $\alpha$ would have to rise to maintain equilibrium, the rapid fall in the activity coefficients of the ions being offset by an increase in their number; the increase in $\alpha$ would then tend to cause an increase in conductivity.

The Debye–Hückel equation for the mean ion activity coefficient is

$$-\log \gamma_\pm=\frac{A(\alpha c)^{1/2}}{1+Ba(\alpha c)^{1/2}} \qquad \text{(C.21)}$$

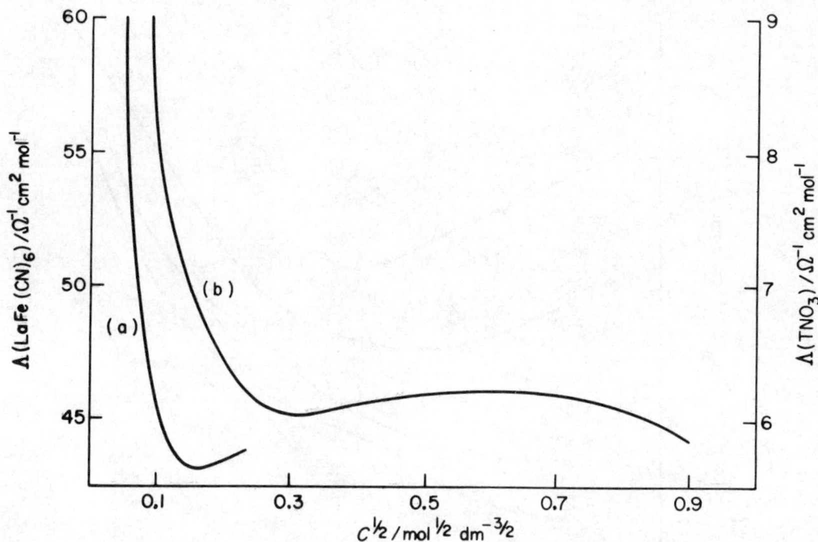


Figure C.12 Curve (a): Tetra-iso-amylammonium nitrate ($TNO_3$) in water–dioxane (14.95% $H_2O$, $\varepsilon = 8.5$). Curve (b): Lanthanum ferricyanide in water.

where $A$ and $B$ are known constants for a given solvent and temperature, and $a$ is the distance of closest approach of the ions. The coefficient $\gamma_{\pm}$ can therefore be evaluated, and for $a = 4 \times 10^{-10}$ m at 298 K equations (C.20) and (C.21) predict a minimum in the degree of dissociation at $\alpha c = 0.005$ mol dm$^{-3}$ for a solvent of dielectric constant 12, and at $\alpha c = 0.001$ mol dm$^{-3}$ for a dielectric constant of 7. This therefore provides an explanation for the findings in non-aqueous solvents.

In water no dissociation minimum would be possible on this basis for a uniunivalent salt, but in $LaFe(CN)_6$, with a valency product of nine, the interionic forces will be comparable with those of a uniunivalent salt in a medium of dielectric constant 9, and the same explanation is applicable (figure C.12). The other instances of minima in water concern substances that form micelles. In a typical example cetylpyridinium bromide contains micelles of the approximate composition $[Cet_{68}Br_{53}]^{15+}$ together with an equivalent number of free $Br^-$ ions. In

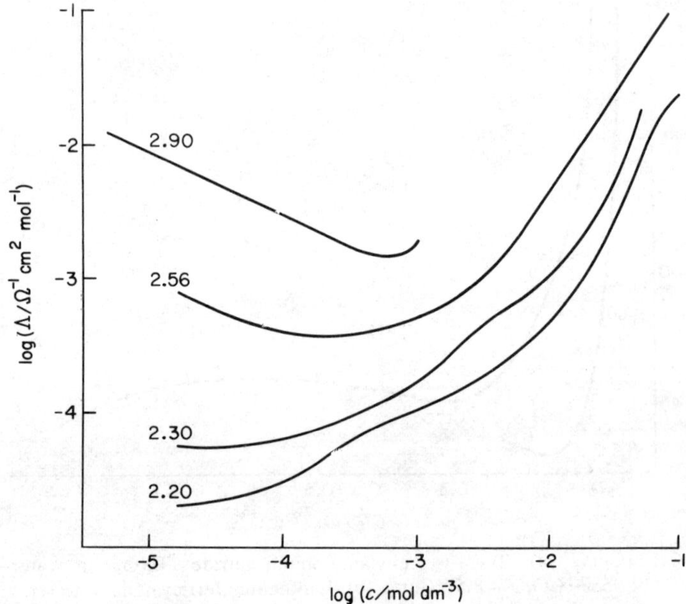


Figure C.13 Conductance of tetra-iso-amylammonium nitrate in dioxane–water mixtures. The dielectric constant of the solvent is shown against curves

this extreme case the quantitative use of equation (C.21) is unjustifiable, but it has been shown that the same explanation, e.g. the process

$$[\mathrm{Cet_{68}Br_{53}}]^{15+} \rightarrow [\mathrm{Cet_{68}Br_{52}}]^{16+} + \mathrm{Br}^-$$

probably applies.

In solvents of very low dielectric constant (figure C.13), conductivity minima are common, and the very low measured $\Lambda$ values may rise with concentration over the whole attainable range. This has been attributed to the formation of triple ions, $+-+$ and $-+-$, followed by further association to quadrupoles or higher aggregates.

*See also* D, H & O.

**Conductance of non-aqueous solutions**

*See* Non-aqueous solutions.

**Conductimetric titration**

For analytical purposes, conductance measurements are most useful for very dilute solutions, or where the electrolyte of interest is a minor constituent, accompanied perhaps by larger amounts of non-electrolyte. The conductimetric method is illustrated in figure C.14(a), which shows the conductance change when HCl is added to NaOH solution. Each addition reacts completely:

$$Na^+ + OH^- (+H^+ + Cl^-) \rightarrow Na^+ + Cl^- (+H_2O)$$

and the number of $OH^-$ ions replaced by the slower $Cl^-$ ions will be proportional to the amount of acid added. After the end-point, the conductance will increase in proportion to the excess of acid added.

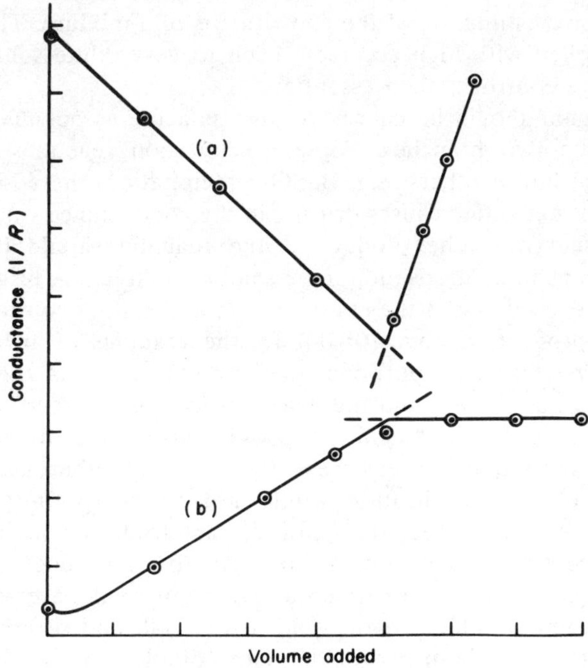


Figure C.14 Curve (a), titration of strong acid with strong base; curve (b), titration of weak acid with weak base

Allowance should be made for the dilution effect. This can be done by multiplying each conductance value, before plotting, by the factor (original volume + added volume)/(original volume); or it may be negligible if a large volume of dilute solution is used in the cell, and a concentrated solution added from a microburette. The corrected values should lie on two straight lines, and it is sufficient to measure three or four points on each branch; the end-point can then be found graphically with considerable accuracy.

This example illustrates one of the advantages of the conductance method; the equivalence point is found graphically, so that irregularities, such as hydrolysis, which interfere with the end-point of a titration do not affect the accuracy. Another advantage, compared with indicator methods, is that it can be used in coloured or turbid solutions. The method also offers special techniques, in suitable cases, for the successive estimation of the constituents of a mixture. The method can be applied with high accuracy, even to very dilute solutions, but temperature control is then essential.

The reagent should be chosen to give as acute as possible an angle between the two branches. Some precipitation reactions are quite satisfactory, but in others, e.g. $BaSO_4$ precipitation, the adsorption of ions on the precipitate causes drifting in the conductance values. When possible, therefore, the product of the reaction should be a non-electrolyte, as in neutralisation, or a soluble complex. A bridge (figure C.8) may be used, and if the resistances $R$ and $r_2$ have constant values, to give a product of, say, 100 000 $\Omega$, the readings of an adjustable resistance $r_1$ will be proportional to the conductance of the solution; these may be recorded, and the whole process made semi-automatic.

The titration of a weak acid (or base) is illustrated in figure C.14(b). The reagent used is a weak base (or acid). In other methods the end-point of such a titration would not be sharp on account of hydrolysis, but this effect does not detract from the accuracy of a conductance titration. A semi-strong acid (p$K$ 2–4) does not give a very satisfactory end-point with either strong or weak base; in this case some ammonia is added to the acid in the cell, and titration is then carried out with sodium hydroxide. This completes the neutralisation of the acid, after which the ammonium ion present is replaced by the sodium ion, with a fall in conductance. Finally the conductance rises as

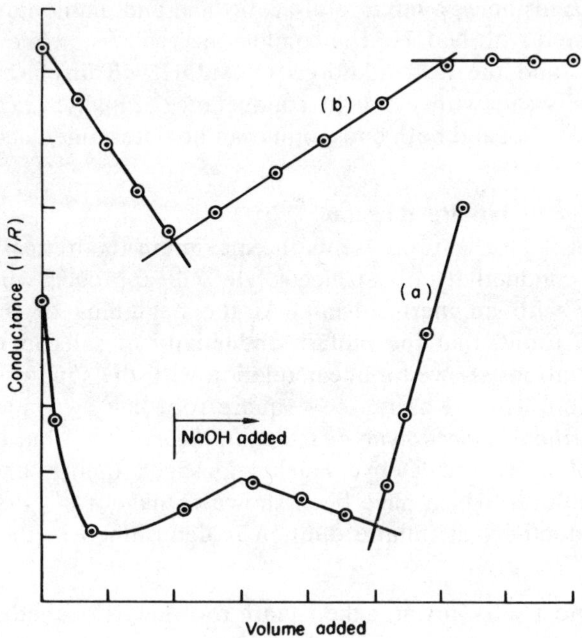


Figure C.15 Curve (a), titration of semi-strong acid; curve (b), titration of HCl–$CH_3CO_2H$ mixture

excess NaOH is added, and at the end-point the amount of NaOH used is equivalent to the acid originally present (figure C.15a).

The components of a mixture of acids can be successively estimated if they differ sufficiently in dissociation constant. Figure C.15(b) shows the titration of a mixture of hydrochloric and acetic acids with ammonia. In the first branch, the dissociation of the acetic acid is suppressed, and the hydrochloric acid is being neutralised; in the second, the acetic acid is being converted into its ammonium salt. Both end-points can be determined accurately, although in the immediate neighbourhood of both there is a slight rounding-off of the graph. A similar shape of curve, with equidistant end-points, has been used to show that an acid is dibasic.

In a 'displacement titration' the salt of a weak acid or base is titrated. For instance, a mixture of acetic acid and ammonium chloride can be titrated with NaOH. The conductance will rise while the acid is neutralised, and the next additions of NaOH will displace the weak base from its salt with a fall in conductance. Finally, excess NaOH gives a sharp rise, and both end-points can be determined accurately.

**Conductivity at infinite dilution**

The symbol $\Lambda^{\infty}$ (or $\Lambda^{\circ}$) represents the maximum theoretical value that the molar conductivity of an electrolyte will approach when diluted indefinitely with an inert solvent. At the beginning of this century Kohlrausch found that the molar conductivity of salts in very dilute aqueous solutions showed a linear relation with the square root of the concentration. This, 'Kohlrausch's square root law', was incompatible with the *Arrhenius electrolytic dissociation theory* (q.v.), but it has since been justified by the Debye–Hückel–Onsager theory of interionic attraction effects, which have been shown to have a $c^{1/2}$ dependence.

The conductivity at infinite dilution is determined in the following ways:

(a) For most salts in water the square root law is obeyed, and the $\Lambda$ values from a concentration of about 0.001 mol dm$^{-3}$ downwards are plotted against $c^{1/2}$, and the straight line obtained is extrapolated to $c = 0$. The slope of the line should be checked against the theoretical slope calculated from Onsager's equation (see *conductance equations*).

(b) $\Lambda$ values over a much wider range of concentration can be utilised in Shedlovsky's equation; this is an empirical extension of the Onsager equation $\Lambda = \Lambda^{\infty} - (a\Lambda^{\infty} + b)c^{1/2}$, which rearranges to give

$$\Lambda^{\infty} = \frac{\Lambda + bc^{1/2}}{1 - ac^{1/2}} \qquad \text{(C.22)}$$

Shedlovsky found that for several salts the right-hand side of equation (C.22) varied linearly with $c$ up to about 0.1 mol dm$^{-3}$. His method of obtaining $\Lambda^{\infty}$ is therefore to plot experimental values in this form against $c$, and to extrapolate the straight line to $c = 0$.

(c) For electrolytes that are not completely dissociated the above methods are inapplicable. Some salts in which the tendency towards incomplete dissociation is not very great may appear to obey the square root law through a compensation of opposing effects, but an extrapolation on this basis will be unreliable. For all such electrolytes it will usually be possible to calculate $\Lambda^{\infty}$ by simple addition of known values for cation and anion, since accurate values for the commoner ions have been obtained by combining conductivity and transport number data (see *molar ionic conductivity*). If this is impossible, e.g. for an acid HX, the $\Lambda^{\infty}$ value for a salt such as KX is found by method (a) or (b), and this is combined with the known values for $H^+$ and $K^+$:

$$\Lambda^{\infty}(HX) = \Lambda^{\infty}(KX) + \Lambda^{\infty}(H^+) - \Lambda^{\infty}(K^+)$$

**Contact potential**

The contact potential is the equilibrium potential difference established when two dissimilar metals $M_1$ and $M_2$ are in contact at a given temperature. It is related to the work function, $\phi$, i.e. the work required to extract electrons from the metals, by

$$V(M_1, M_2) = \phi(M_1) - \phi(M_2)$$

$\phi(\mathrm{Pt}) = 4.52\ \mathrm{eV} > \phi(\mathrm{Cd}) = 4.00\ \mathrm{eV}$, so that less energy is required to remove an electron from Cd than from Pt, and more energy is released when the electron drops into the Pt lattice than when it drops into the Cd lattice. Transfer of an electron from Cd to Pt at the same potential results in the spontaneous release of 0.52 eV of energy, i.e. the contact potential of Cd relative to Pt is −0.52 eV. This net transfer of electrons makes the Pt negative with respect to Cd. Thus, at equilibrium, when electron flow is the same in both directions, the equilibrium potential is the contact potential. If a third metal is introduced between the other two, the potential difference between the two end metals will be the same as if they were directly in contact.

One method of measuring contact potentials is to determine the potential difference between the plates of the metals in contact using a quadrant electrometer.

Values of contact potentials listed for different metals are with reference to a standard metal, usually Pt. Contact potentials are extremely sensitive to surface contamination (e.g. oxide formation, adsorption of polar vapours); for this reason they are of use in adsorption phenomena.

**Copper coulometer**

The copper coulometer, often called the 'copper voltameter', is a convenient way of measuring quantities of electricity and is quite accurate for most purposes. For comparatively small currents, such as are used in transport number experiments, the arrangement shown in figure C.16 is satisfactory. The small beaker contains a solution made up in the following proportions: hydrated copper sulphate 15 g, sulphuric acid 5 g, ethyl alcohol 5 g, water 100 g. The two electrodes are of copper foil thoroughly cleaned with abrasive powder, and have small holes punched in the tags to enable them to hang from the copper leads. When the experiment is to begin, the electrodes are placed in the solution, and a pink deposit of copper forms on the cathode while copper dissolves from the anode. The current can be broken by lifting the cathode off its hook; it is then washed successively with water and alcohol, dried over a small flame and weighed. The gain in weight gives the number of coulombs passed (0.3295 mg,

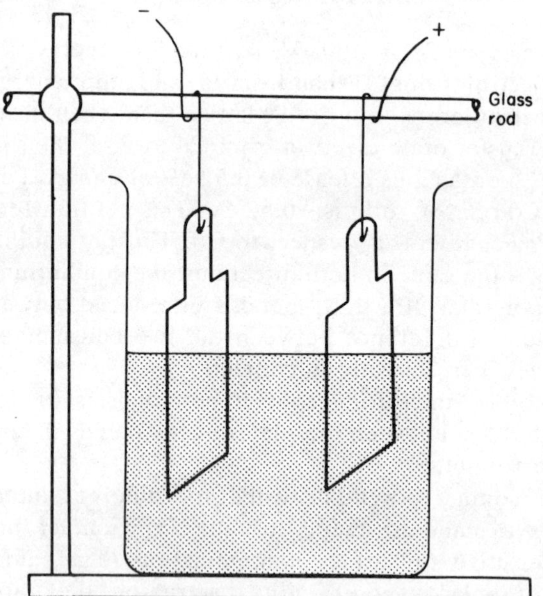


Figure C.16 Copper coulometer for small currents

the electrochemical equivalent of copper, is deposited by 1 C). For heavier currents more substantial coulometers of the same type are used; these often have two anode plates, wired in parallel, on either side of the cathode.

**Copper, electrometallurgy**

Copper can form two series of salts, and thermodynamic data give the following equilibrium potentials:

$$E(Cu^{2+}, Cu) = +0.337 + (RT/2F) \ln a(Cu^{2+})$$

and

$$E(Cu^{+}, Cu) = +0.521 + (RT/F) \ln a(Cu^{+})$$

When a solution containing copper ions is in contact with a copper electrode, the system will move spontaneously towards a position of equilibrium through a transfer of electrons between the species $Cu^{2+}$, $Cu^{+}$ and $Cu^{0}$. A potential $E$ is thus established which satisfies both equations, and

$$0.337 + (RT/2F) \ln a(Cu^{2+}) = 0.521 + (RT/2F) \ln a^{2}(Cu^{+})$$

or

$$0.184 = (RT/2F) \ln a(Cu^{2+})/a^{2}(Cu^{+}) = (RT/2F) \ln (1/K) \qquad \text{(C.23)}$$

where $K$ is the equilibrium constant for the reaction $Cu^{2+} + Cu \rightleftharpoons 2Cu^{+}$, which comes to $6.4 \times 10^{-7}$. If the solution is of unit activity with respect to copper(II) ion, equation (C.23) demands that the activity of copper(I) ion should be given by $0.184/0.029 = -2 \log a(Cu^{+})$, or $a(Cu^{+}) = 8 \times 10^{-4}$ mol dm$^{-3}$. Copper dissolving under equilibrium conditions at a copper anode would give the two ions in the ratio required by the equilibrium constant.

Electrolysis is used for extracting copper from poor ores, and for reclaiming scrap copper. The copper is brought into solution as copper(II) sulphate, and this is electrolysed with a cathode of sheet copper and a lead-alloy anode. Oxygen is liberated at the anode and sulphuric acid accumulates in the electrolyte. When the electrolyte is exhausted, the sulphuric acid is used to bring more copper into solution.

Owing to the demand for very pure copper, electrolytic refining is practised on a very large scale. The cathodes are thin sheets of copper and the anodes blocks of the impure metal, and the electrolyte consists of copper(II) sulphate and free sulphuric acid; the presence of the latter improves the conductance and has other advantages. The current density is kept below the value at which hydrogen might be simultaneously discharged, but there is an appreciable *overpotential* (q.v.) at both electrodes. One result of this is that copper(I) ions are formed at the anode at concentrations greater than that corresponding to equilibrium in the bulk of the solution. As the ions diffuse away from the anode, therefore, the process $2Cu^+ \rightarrow Cu^{2+} + Cu$ will tend to occur, leading to the deposition of powdery copper in the bath. The acid electrolyte enables this to be brought back into solution by the reaction

$$Cu + H_2SO_4 + \tfrac{1}{2}O_2 \rightarrow CuSO_4 + H_2O$$

Of the impurities in the anode, the silver and other noble metals pass into the anode sludge, and any selenium, tellurium, arsenic, antimony and bismuth are also precipitated as insoluble compounds. Other impurities, especially nickel and iron, pass into solution and gradually accumulate in the electrolyte, which has eventually to be renewed.

A bipolar system is sometimes used, in which a number of plates of impure copper, not connected to the electricity supply, are spaced out between the cathode and anode. Each acts as a *bipolar electrode* (q.v.), copper dissolving from one face while pure copper is deposited on the other, so that they eventually consist of pure metal. The energy requirement is much less for this than for the simple arrangement, but there is some leakage of current around the plates. Strict control is also necessary, as an insufficient passage of current would leave layers of impure metal, while any excess of current would be entirely wasted.

**Copper oxide cell**

Copper oxide cells are reliable and fairly cheap, and are widely used as wet cells. The cathodes are porous plates of copper oxide mixed with a suitable binder, the anodes are of zinc; and the electrolyte is a concentrated sodium hydroxide solution. Zinc ions from the anode react to give zincate ions, and copper oxide is reduced at the cathode:

$$Zn + 3OH^- \rightarrow HZnO_2^- + H_2O + 2e$$

$$CuO + 2e + H_2O \rightarrow Cu + 2OH^-$$

so that the over-all reaction is

$$Zn + CuO + OH^- \rightarrow Cu + HZnO_2^-$$

The cell gives a steady voltage of 0.6–0.7 V, which shows little falling off during the discharge of the cell.

**Corrosion**

The electrochemical aspects of corrosion can be illustrated by considering a piece of zinc containing inclusions of a more positive metal such as copper. If this is wetted by an aqueous solution containing the metal ions, the *Daniell cell* (q.v.) reaction will occur by a process of *internal electrolysis* (q.v.). Zinc ions will pass into solution, the electrons released will move through the metal to the more positive regions where copper ions will be discharged by them, and so the corrosion of the zinc will continue. Some practical examples of corrosion occur in this way, when two different metals are in electrical contact with a surface film of moisture to provide the electrolyte. But the presence of a second metal is not necessary; local regions of relatively positive or negative potential may result from differences in the activities of surface atoms; the potential of a cold-worked metal is usually slightly more negative than that of an annealed metal, and in the same piece of metal, grain boundaries or two different crystal faces can provide the two poles of a cell.

Heterogeneities in the metal surface are not necessary, however, for corrosion to occur. Electrolytic action requires an anode (at which electrons are released) in electrical contact with a cathode (at which they are consumed by chemical action) and an electrolyte to transport the current through the moisture film or other liquid phase. The anode reaction in corrosion is the dissolution of metal. The cathode reaction need not be the discharge of a metal ion, as in the first example considered above. In aqueous solutions the possible reactions include reduction of oxygen,

$$O_2 + 4H^+ + 4e \rightarrow 2H_2O$$

or

$$\tfrac{1}{2}O_2 + H_2O + 2e \rightarrow 2OH^-$$

and hydrogen evolution,

$$2H^+ + 2e \rightarrow H_2$$

and reducible impurities may provide others. So long as a pure metal surface is at a potential more positive than its standard electrode potential and more negative than the equilibrium potential of any one of these possible cathodic processes, both cathodic and anodic reactions can occur spontaneously, and the metal will corrode at random points over the surface. Control of the rate of this corrosion may be achieved by eliminating the cathodic reaction which takes place most readily.

In addition to these effects, corrosion can be caused through local variations in the solution concentrations, by a type of *concentration cell* (q.v.) effect. A common example of this is illustrated in figure C.17 for a steel structure immersed in sea-water. The concentration of oxygen decreases with increasing depth, and so will the equilibrium potential for oxygen reduction. As a consequence, the metal near the surface will act as a cathode, and metal dissolution will occur in the parts

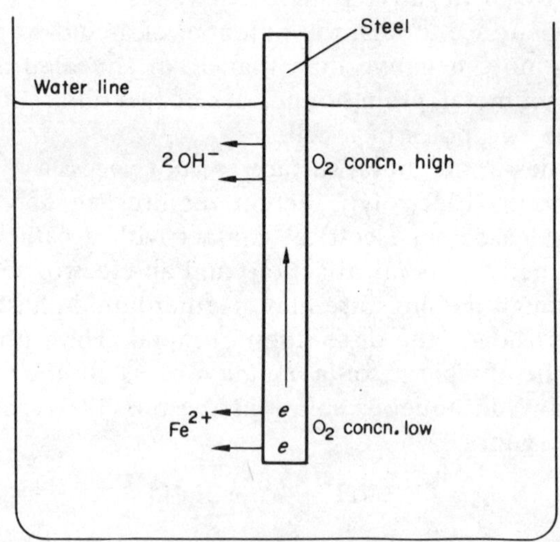


Figure C.17 Corrosion in region remote from water-line

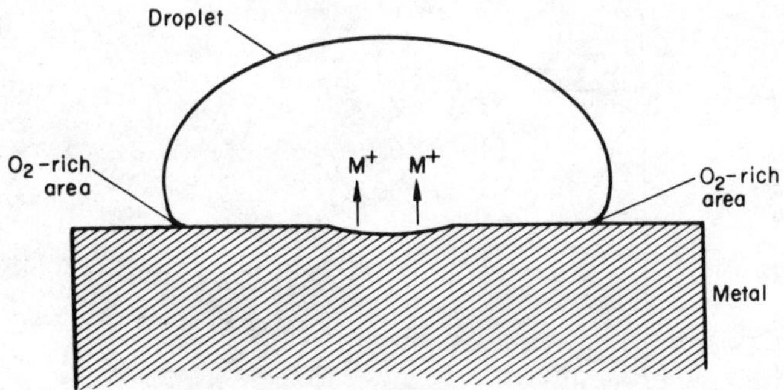


Figure C.18 Differential aeration causing corrosion in water droplet

further from the water-line. The same effect can occur with buried pipes if the availability of oxygen varies from place to place. A further illustration of the effect is given in figure C.18, where an isolated droplet of moisture is shown on the surface of a corrodible metal. Oxygen is more readily available at the periphery of the drop and will be reduced here, while metal ions will dissolve at the centre of the drop and eventually precipitate, in most cases, as hydroxide.

A semi-quantitative treatment of corrosion rates becomes possible if the current–potential curves are known for the dissolution reaction and the possible cathodic reactions. Figure C.19 shows the *exchange current density* (q.v.) and the Tafel lines for zinc electrodes in acid solution. The electrode will be in equilibrium with zinc ions at a potential $E^{\ominus}(Zn^{2+}, Zn) = -0.76$ V, and for the process

$$Zn \underset{j_0}{\overset{j_0}{\rightleftharpoons}} Zn^{2+} + 2e$$

the point on the diagram with intercepts $E^{\ominus}(Zn^{2+}, Zn)$, log $j_0(Zn)$, will be the origin of the Tafel line. At more positive potentials, zinc will pass into solution at a finite rate, the current density being shown by curve (a) of figure C.19. The equilibrium potential for hydrogen discharge will occur at a finite rate for potentials more negative than $E^{\ominus}(H^+, H_2)$. The Tafel lines for a cathodic and an anodic reaction must tend to intersect, as they do in figure C.19 at $E_{corr}$. At this point the

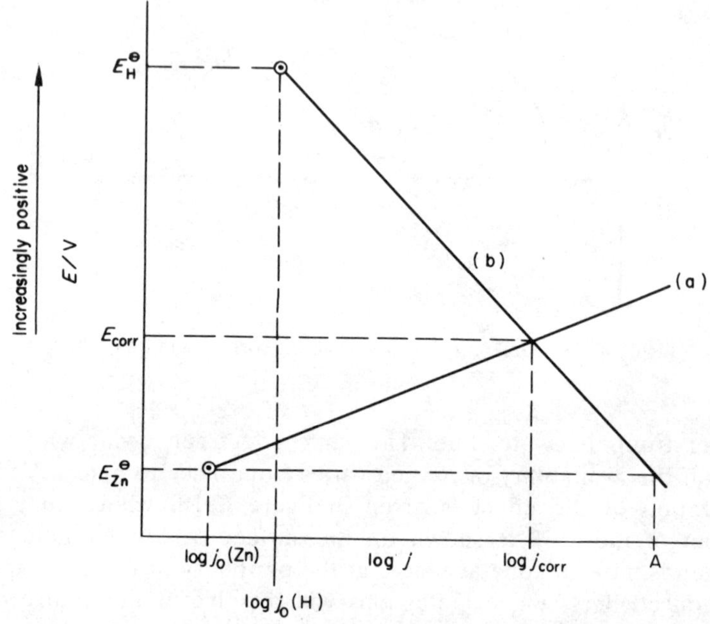


Figure C.19 Derivation of the corrosion potential

current density for both reactions is the same; the electrons made available by the dissolution of zinc ions are absorbed at an equal rate by the discharge of hydrogen, and a steady state is established at $E_{corr}$, the corrosion potential. The theoretical rate of corrosion is measured by the value $j_{corr}$. A diagram of this kind can be constructed if the constituents of the solution and the probable cathode reaction are known, and if the $E$–$j$ relationship for this has been determined.

Figure C.20 illustrates some of the influences that can affect the rate of corrosion of a metal. Pure zinc (curve a), amalgamated zinc (curve b) and zinc containing inclusions of platinum (curve c) are compared. The presence of platinum greatly increases the corrosion rate, partly because of the high $j_0$ value for hydrogen at this low overpotential metal, and partly because of the change in slope of the Tafel line (explained by a different mechanism for the *hydrogen evolution reaction* q.v.).

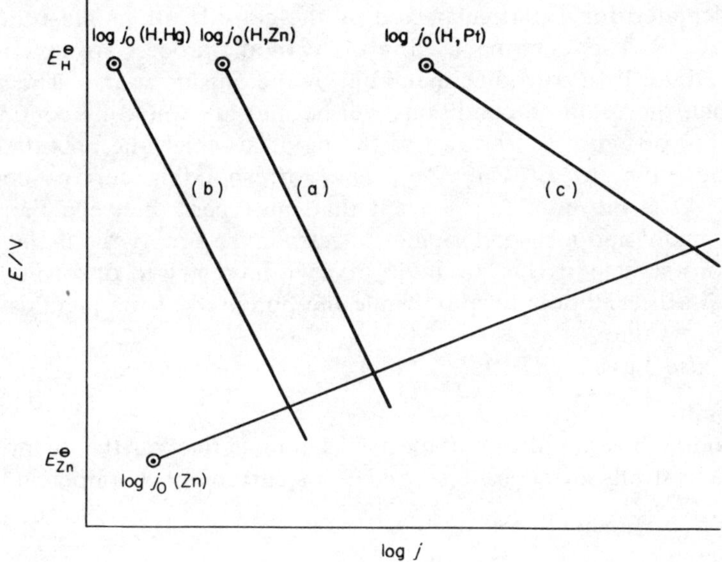


Figure C.20 Corrosion rates and potentials at zinc electrodes

*Prevention of corrosion*

The possibilities include control of the environment (perhaps by humidity control, de-aeration, or addition of inhibitors which retard the cathodic process) and control of the surface (by surface coatings, electroplating or producing surface layers of high resistance, as in the phosphate treatment of steels). There are also two more purely electrochemical methods.

One is to connect the metal to be protected to a more active metal, i.e. one with a more negative $E^{\ominus}$ value. The anodic reaction will then occur preferentially at the surface of this, which will provide protection while itself gradually going into solution. This is an example of sacrificial protection.

The second method is to impose a potential from an outside source on the metal to be protected, to ensure that the potential at its surface is never more positive than its $E^{\ominus}$ value. Under these conditions the metal will not corrode. The energy requirements for this method can be calculated for a particular case by the considerations illustrated in figure C.19. The electrons used in the cathodic process are now being supplied not by a corroding metal but by the outside source. The value at which the required conditions will be met are found by continuing curve (b) of figure C.19 down to the point at which the potential has become equal to $E^{\ominus}(Zn^{2+}, Zn)$. The corresponding current density (point A) is the minimum current that must pass (between the protected metal and a secondary inert electrode) to supply the metal with electrons sufficiently fast to nullify its own tendency to dissolve; these maintain the cathodic reaction while the potential of the metal is held at its $E^{\ominus}$ value.

*See also* Passivity; E, Fr.

**Coulomb**

The coulomb is the SI unit of electric charge, defined as the quantity of electricity transported in 1 second by a current of 1 ampere (1 C = 1 A s).

**Coulometer**

The quantity of electricity passed through a cell may be determined by measuring the current as a function of time, and determining the area under the current–time curve. A calibrated galvanometer with a short response time is used. Alternatively a chemical coulometer is commonly employed. This consists of an electrolytic cell in series with the experimental cell, the same amount of electricity therefore passing through both; the chemical reaction at the cathode or anode (or both) of the coulometer must occur with 100% current efficiency, and should be easily and accurately estimated. The deposition of silver at the cathode of a *silver coulometer* (q.v.), the dissolution of silver from a silver anode (see *Faraday constant*) and the reaction $2e + I_2 \rightleftharpoons 2I^-$ of the *iodine coulometer* (q.v.) all satisfy these conditions, and another common and convenient device is the *copper coulometer* (q.v.).

When the quantity of electricity is small, a colorimetric method of estimation will be more sensitive. This has been applied to the iodine formed in an iodine coulometer, and to silver dissolved from a silver anode (estimated as silver dithizonate). Volumetric analysis has also been applied: for instance, when water is electrolysed, the reaction in the cathode compartment is $e + H_2O \rightarrow \frac{1}{2}H_2 + OH^-$, and the alkali may be titrated with standard acid; 1 $cm^3$ of 0.01 mol $dm^{-3}$ acid will be equivalent to 0.9649 C. Finally, the water reaction can be used gasometrically. In a cell where the cathode and anode portions can mix the reaction $2H_2O \rightarrow 2H_2 + O_2$ requires $4F$, and from the total volume of gas evolved (from a saturated solution) quantities of electricity down to 10 C can be accurately measured (0.1791 $cm^3$ at s.t.p. $\equiv$ 1 C).

**Coulometry**

This method of analysis depends on Faraday's law of proportionality (see *Faraday's laws*) between the quantity of electricity passing in an electrolytic process and the amount of chemical action. Hence, a measurement of the quantity of electricity, or the time for which a constant current passes, can be used for analytical determinations. The method requires carefully controlled conditions, and two essentials are that the selected process must go with 100% current efficiency, and that its end-point must be accurately indicated in some way. It can be used for the rapid determination of very small quantities in suitable cases.

One application is in determining the thickness of an electrodeposit. A known small area is used as anode, the rest of the article being blocked off with a suitable coating. The electrodeposit is dissolved at an anode potential which leaves the base metal unattacked, and a sudden rise in this potential indicates the end of the reaction. The quantity of electricity used may be given by an accurate *coulometer* (q.v.) in series with the cell.

A constant current, controlled by an amperostat, can also be used. Small quantities of arsenite have been estimated by electrolysing the solution, which contains $H_2SO_4$ and NaBr, between Pt electrodes at a constant current of 1–10 mA. A diaphragm cell is used, and in the anode compartment the bromine formed immediately oxidises $AsO_3^{3-}$ to $AsO_4^{4-}$. The end-point is marked by the appearance of free $Br_2$, and one way of recognising this is by a secondary circuit consisting of two Pt electrodes maintained at a voltage difference of 0.2 V. This is insufficient to oxidise $AsO_3^{3-}$ directly, but enables the free $Br_2$ to react: $\frac{1}{2}Br_2 + e \rightarrow Br^-$. Hence, a sudden rise in current in this circuit indicates the end of the reaction, and can be used to interrupt the main circuit and actuate a time recorder.

There are limited applications also for 'coulogravimetry', in which both the gain (or loss) in weight of an electrode and the current responsible are measured. Suppose that two bivalent metals of relative atomic masses $A_1$ and $A_2$ are codeposited on a cathode. The gain in weight is given by $w = w_1 + w_2$, where these are the weights of the two metals deposited. The current requirement, however, is given by

$$Q = 2 \times 96\,500(w_1/A_1 + w_2/A_2)$$

where $Q$ is the number of coulombs used; these simultaneous equations enable the ratio to be determined in which the two metals are deposited.

*See also* L.

**Current efficiency**

In an electrolytic process the current efficiency is the percentage of current utilised in the process under consideration. Suppose, for instance, that when 0.01 Faraday of electricity is passed through a silver salt solution 1.0540 g of silver are deposited on the cathode. Theoretically $A/100 = 1.0788$ g should have been deposited, where $A$ is the relative atomic mass of silver. The current efficiency for silver deposition is therefore $105.4/1.0788 = 97.7\%$ under these conditions, and 2.3% of the current has been used in side reactions.

The 'current efficiency' quoted for a storage battery, such as a lead accumulator, is

$$100 \times \frac{\text{charge delivered during discharge}}{\text{charge used in charging process}}$$

and is really an ampere-hour efficiency.

# D

**Daniell cell**

The Daniell cell is a simple primary cell consisting of a zinc electrode partially immersed in a zinc sulphate solution and a copper electrode partially immersed in a copper sulphate solution; the two solutions are

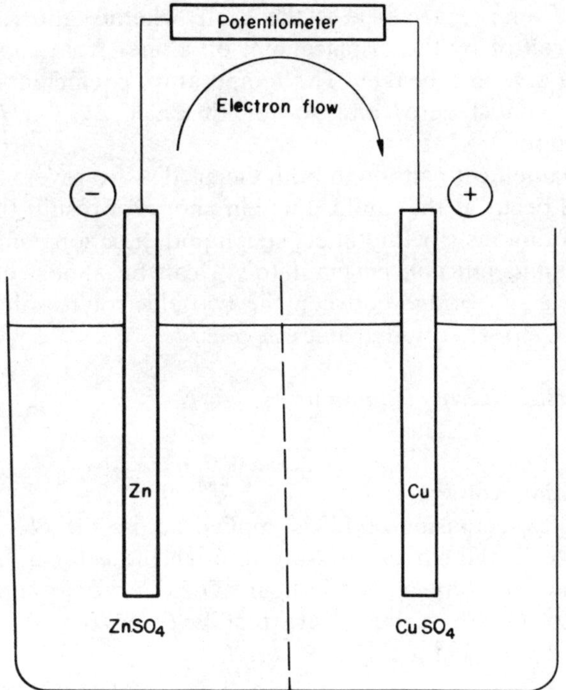


Figure D.1 Daniell cell

separated by a permeable membrane or sintered glass partition (figure D.1).

When the external circuit is completed, positive current flows from left to right inside the cell, and from right to left outside. Electron flow through the external circuit is left to right. The current is carried inside the cell at the $Zn \mid Zn^{2+}$ boundary by $Zn^{2+}$ ions passing into solution, at the $Zn^{2+} \vdots Cu^{2+}$ interface partly by $Zn^{2+}$ ions and partly by $SO_4^{2-}$ ions, and by $Cu^{2+}$ ions which are deposited on the copper electrode. The over-all reaction producing the chemical change and, hence, the electrical energy is

$$Zn + Cu^{2+} \rightarrow Zn^{2+} + Cu$$

$E(298\text{ K}) = 1.1\text{ V}$; hence, $\Delta G = -2 \times 96\,487 \times 1.1\text{ J} = -212\text{ kJ}$.

This free energy decrease is the same whether the reaction takes place in a cell or by the displacement of copper from copper sulphate solution by zinc in a beaker. The temperature coefficient of the e.m.f. ($\partial E/\partial T$) is almost zero, and so for this cell $\Delta G = \Delta H$ to a good approximation.

In the Daniell cell, although both electrodes are reversible, diffusion of ions will occur at the liquid junction and, as a result, the e.m.f. will change continuously with time. The liquid junction, and the corresponding *liquid junction potential* (q.v.), can be almost eliminated by interposing a salt bridge between the two electrolyte solutions.

*See also* Cell; Thermodynamics of cells.

**Debye–Hückel activity equation**
*See* J.

**Decomposition voltage**
If a gradually increasing e.m.f. is applied across the electrodes of an electrolytic cell, the current passing through the cell can be plotted to give curves of the type shown in figure D.2. Curve (a) applies to cases where the electrode reactions are readily reversible. At first, a very

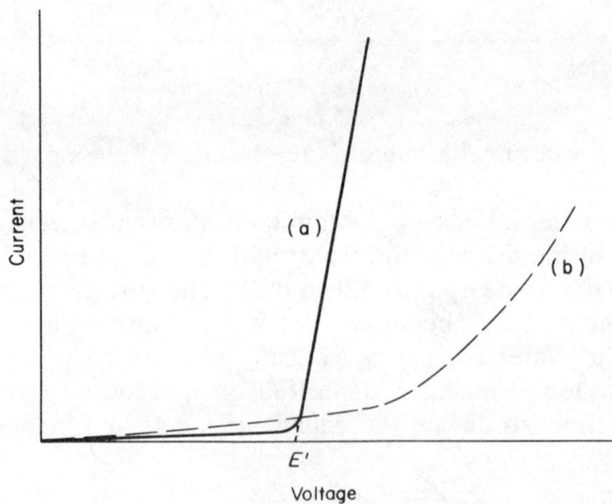


Figure D.2 Current–voltage curves

small current, called the residual current, passes through the cell. This must consist partly of the very minute current needed to charge the electrodes and set up the gradually increasing electrical double layer at each electrode surface; this is called a 'non-faradaic current'. The rest of the residual current must be a 'faradaic current', i.e. one that is due to electrolytic processes, such as the oxidation or reduction of traces of impurities, or the maintenance of equilibrium concentrations at the electrode against losses by diffusion. Eventually the current turns steeply upwards. The applied voltage is now sufficient to sustain the main electrolytic reaction, and by extrapolating the second branch of the curve back to zero current, an experimental value $E'$ for the decomposition voltage is obtained. For a case such as this, in which the electrode reactions occur readily, a very small increase of the e.m.f. beyond this value is theoretically sufficient to maintain large currents, and the excess voltage is mainly used in overcoming the resistance of the solution. The equation of the straight line is then $E = E' + IR$, and the slope is governed mainly by $R$, the resistance of the solution.

Curve (b) (figure D.2) is obtained if the electrode reactions take place less readily, e.g. if there is a slow step in the mechanism of the over-all process. In this case appreciable currents will only be observed at voltages higher than the theoretical $E'$ values, and the difference $E - E'$, the *overvoltage* (q.v.) will be the higher the higher the current passed. In this case a value $E'$ cannot be obtained by extrapolation.

For any electrolytic process a theoretical value for the decomposition voltage can be obtained from thermodynamic data, but in a closer analysis it will clearly be better to study the two electrode reactions separately. This can be done with the arrangement shown in figure D.3. The electrolysis current is best provided by a high-tension battery, with a resistance $R$ in series with the cell. $R$ is so high that minor changes in the cell are insignificant, and a constant current, measured by the ammeter A, can be maintained. The electrodes are thus 'polarised' to an extent corresponding to the measured current, and in this condition they can be combined in turn with a reference electrode. The cell thus formed is balanced on the potentiometer, and its e.m.f. measured with no current passing through the measuring circuit. The connection to the reference electrode is by means of a 'Luggin capillary'. This is a tube with its tip brought very close to the surface of the

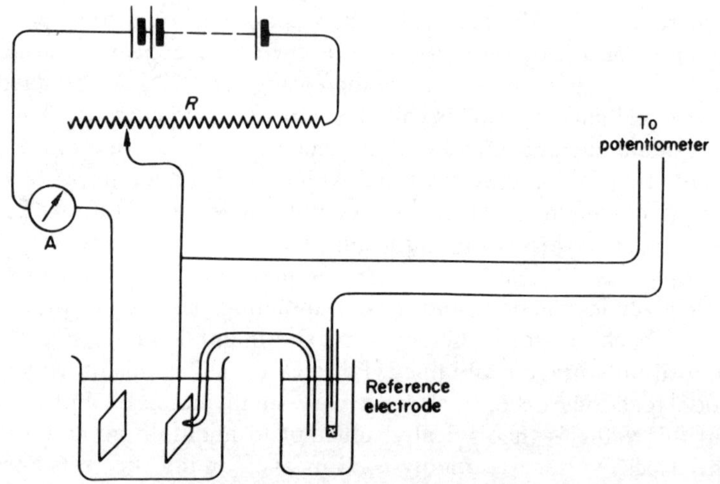


Figure D.3 Measurement of cathode and anode potentials

working electrode. Thus, no part of the *IR* drop in the electrolytic cell is included in the measurement, and with this complication eliminated the measured e.m.f. is the sum of the known potential of the reference electrode and the potential of the electrode under study. Curves such as those in figure D.2 can thus be plotted for cathode and anode separately and, where possible, extrapolated to give the decomposition potential. Cathode and anode potentials are linked by the requirement that the current passing is the same at both electrodes, and from the two current potential curves the potentials corresponding to a given current can be read off, and combined to give the total potential drop. The amount by which the electrode potential exceeds the reversible potential for the electrode concerned is the cathode or anode *over-potential* (q.v.).

**Demineralisation of water**

An electrolytic method of purifying brackish water uses cells containing alternate cation-exchange and anion-exchange membranes, as in figure D.4. The cation-exchange membrane is made from a polysulphonate resin, which contains a high concentration of negatively

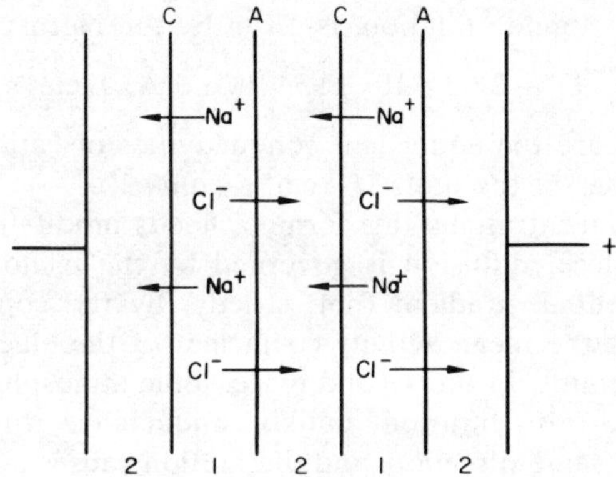


Figure D.4 Demineralisation cell

charged groups fixed to the resin network. Owing to the Donnan effect (see *Donnan membrane equilibrium*‡), anions are virtually excluded from the pores of the membrane, and the transport of electricity through it is confined to the movement of cations. In the same way, anion-exchange membranes transport anions only. When an electric field is applied across an assembly of such cells, desalted water flows from alternate channels (marked 1), and the ions are carried away in the channels 2. In practice, hundreds of closely spaced membranes are used in a multicompartment cell.

**Dielectric constant**
Also known as relative permittivity. *See* Permittivity.

**Diffusion in electrolyte solutions**
Consider the diffusion of a completely dissociated uniunivalent electrolyte. Cations and anions will begin to move independently down the concentration gradient, but unless they have equal mobilities, the faster ion will tend to get ahead of the other. This will give rise to an electric field which will assist the diffusion of the slower ion but will retard the faster ion, with the result that both ions move at the same speed, and electroneutrality is preserved. The electrolyte thus diffuses at a rate determined by the mobilities of the two ions, and its diffusion

coefficient, for infinite dilution, is given by the Nernst equation

$$D^{\circ} = 2.66 \times 10^{-7}\{\Lambda_c^{\infty}\Lambda_a^{\infty}/(\Lambda_c^{\infty} + \Lambda_a^{\infty})\} \text{ cm}^2 \text{ s}^{-1}$$

where $\Lambda_c^{\infty}$, $\Lambda_a^{\infty}$ are the equivalent conductivities of cation and anion at infinite dilution (in the units $\Omega^{-1}$ cm$^2$ g-equiv.$^{-1}$).

At finite concentrations this formula needs modifying in two ways. In the first place, diffusion is governed by the osmotic pressure, or chemical potential, gradient (not, strictly, by the concentration gradient), so that the mean activity coefficient of the electrolyte must be taken into account. In the second place, ionic atmosphere effects must be allowed for. In diffusion, unlike conductance, the two ions are moving in the same direction, and the motion causes no disturbance of the symmetries of the ionic atmospheres; there is therefore no relaxation effect. There is a small electrophoretic effect, however, the magnitude of which for dilute solutions has been worked out by Onsager, and the most accurate measurements support the extended formula based on these corrections.

**Digital voltmeter**

The digital voltmeter is a rugged precision instrument with electronic counting circuits used to give a digital display of input voltage, in the range 1 mV to 1000 V. A common circuit arrangement employs a ramp-function generator in which a voltage periodically starts from zero and increases at a uniform and accurately controlled rate. An electronic counter counts the number of cycles of a precision oscillator between the starting instant and the instant when the ramp voltage equals the input voltage; the count is then proportional to the applied voltage. The voltage is displayed visually on neon indicator tubes, usually covering five decades. All instruments have a built-in standard cell, and most have automatic positioning of the decimal place and an indication of positive or negative polarity.

Digital voltmeters can be used to automatically track input signal variations, then after a change of input voltage the reading will remain stationary until the next change causes the instrument to trip. The accuracy of commercial instruments varies in the range about 0.001% full-scale deflection, ±0.0025% of reading. Such instruments are used for test purposes in electronics, for general-purpose accurate voltage measurements in electrochemistry and as a built-in digital indicator for control and monitoring panels in industry. Since these instruments have high input impedances (>100 MΩ), they can, with advantage, be used to replace the *potentiometer* (q.v.) in the measurement of the e.m.f. of a *reversible galvanic cell* (q.v.).

**Dissociation constant**
*See* Conductance of aqueous solutions; J.

**Dorn effect**
The Dorn effect is one of the *electrokinetic effects* (q.v.), and is the converse of *electrophoresis* (q.v.). Charged particles falling through a stationary liquid set up a potential difference in the liquid column, opposing the motion of the particles.

Large effects, of the order of 100 V, have been measured, but accurate results are more difficult to obtain than for the other related effects. The effect, also called the sedimentation potential, must be taken into account—for example, in measurements in the ultracentrifuge.

**Double layer**
*See* Electrical double layer.

**Dry cell**
The majority of dry cells are based on the Leclanché cell, which consists of an ammonium chloride solution into which dips a zinc rod as anode. The cathode, in a porous container, consists of a carbon rod packed round with powdered manganese dioxide and carbon. The anode reaction is

$$Zn \rightarrow Zn^{2+} + 2e$$

and at the cathode

$$MnO_2 + H_2O + e \rightarrow MnO(OH) + OH^-$$

The zinc ions react with the alkaline ammonium chloride and a zinc ammine crystallises out:

$$Zn^{2+} + 2OH^- + 2NH_4^+ \rightarrow Zn(NH_3)_2^{2+} + 2H_2O$$

so that the over-all reaction is

$$Zn + 2MnO_2 + 2NH_4Cl \rightarrow Zn(NH_3)_2Cl_2 + 2MnO(OH)$$

The cathode reaction may consist of a primary electrochemical step

$$MnO_2 + 4H^+ + 2e \rightarrow Mn^{2+} + 2H_2O$$

followed by chemical steps resulting in the reaction

$$Mn^{2+} + MnO_2 + 2OH^- \rightarrow 2MnO(OH)$$

In the common dry cell the electrolyte is a concentrated ammonium chloride solution containing some zinc chloride and a small amount of mercury(II) chloride. The latter is reduced to mercury at the zinc surface, which reduces the *corrosion* (q.v.) rate and adds to the shelf life of the cell. The usual form of the cell is a zinc container, acting as anode, next to which is a layer of the electrolyte mixture made into a gelled paste by addition of starch or flour. Inside this is the $MnO_2$–C mixture surrounding a central carbon rod; the latter is fitted with a metal cap to act as positive terminal, and the cell is sealed with a bitumen layer.

A similarly constructed cell uses magnesium in place of zinc, with a magnesium bromide electrolyte. Magnesium has a reversible potential 1.5 V more negative than that of zinc, but it is a mixed potential that controls the electrode in an aqueous medium, and the cell gives about 1.9 V compared with about 1.6 V for the Leclanché cell.

Alkaline-manganese cells have a lower internal resistance and give a better performance than the ordinary dry cell but are more expensive. The electrolyte is a potassium hydroxide solution. The primary electrode reactions are as before, but the zinc ions react with the hydroxide ions, resulting in precipitation of zinc hydroxide and oxide, so that the over-all cell reaction is

$$Zn + 2MnO_2 + 2H_2O \rightarrow 2MnO(OH) + Zn(OH)_2 \quad [\rightleftharpoons ZnO + H_2O]$$

*See also* Primary cell; Mi, P.

# E

**Electrical double layer**
At any boundary between two phases the intermolecular forces will be different from those in the interior of either phase; consequently the concentrations of any mobile species are likely to differ in this region from those in the bulk phases, and any dipolar molecules in the region may be oriented by surface forces. A result to be expected is the setting up of an electrical double layer at the surface, one side of which carries a positive charge and the other an equal negative charge.

The separation of charges may arise in various ways. At a solid–electrolyte solution interface there is likely to be preferential adsorption of one ion or another on the solid surface, with a corresponding excess of the counterion in the adjacent solution. (Even at the air surface of an electrolyte solution there is an electrical separation, as most anions tend to approach the surface more closely than cations.) Above all, the effect will be important where a charged species is able to pass across the boundary from one phase to the other, as in the ionisation of surface groups of a colloidal particle, or at the interface salt crystal–solution, or at a metal electrode.

The simplest model of an electrical double layer is one in which a charge $+Q$ is uniformly distributed over a plane of surface area $A$ and is separated by a distance $d$ from a similar plane with a charge $-Q$. Writing $\sigma = Q/A$ for the surface charge density, we obtain the potential difference across the layer from the formula for a parallel-plate capacitor

$$V = 4\pi\sigma d/\varepsilon \qquad \text{(E.1)}$$

The *capacitance* (q.v.) of the layer is given by

$$C = \sigma/V = \varepsilon/4\pi d \qquad \text{(E.2)}$$

where $C$ is in F cm$^{-2}$ if $V$ is in volts and $\sigma$ in C cm$^{-2}$. This model is called the 'Helmholtz double layer'. In 1905 Gouy pointed out that the Helmholtz double layer cannot apply where one side of the layer consists of mobile ions whose thermal energy would oppose their ordering by electrostatic forces. An equilibrium must result in which

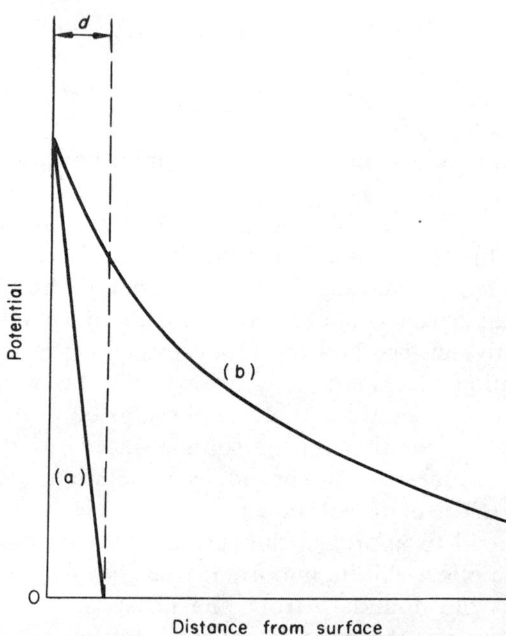


Figure E.1 Electrical double layer: (a) Helmholtz double layer; (b) diffuse double layer

the excess of cation or anion will be at a maximum close to the surface, but will diminish gradually at increasing distances. The quantitative theory is exactly similar to the Debye–Hückel treatment of the atmosphere round a charged ion. This is the Gouy–Chapman model of a 'diffuse double layer'; it is compared with the Helmholtz model in figure E.1. The diffuse double layer is electrically equivalent to a charge, equal and opposite in sign to the net charge on the fixed side of the layer, spread out at a distance $\kappa^{-1}$ from the surface, where $\kappa$ is the Debye–Hückel function and depends on the concentrations and valencies of all the ions present, and $\kappa^{-1}$ is the 'thickness of the ionic atmosphere'. Hence, the diffuse double layer can be treated as a parallel plate capacitor in estimating its effects, with $\kappa^{-1}$ replacing $d$ in equations (E.1) and (E.2).

Later, in 1924, Stern proposed a synthesis of these two models. It is reasonable to suppose that some of the charge carriers in the solution will be strongly bound by surface forces, while the remainder will distribute themselves in a diffuse double layer. The two together will balance the surface charge; if this is $+e_0$, and the charge on the fixed layer is $-e$, there will be a net charge of $-(e_0 - e)$ distributed in the diffuse double layer. There is a possibility also that strong adsorptive forces may lead to a charge on the solution side of the surface greater than the charge on the solid. This will lead to a reversal of sign of the charge of the diffuse double layer, the charges now being, respectively,

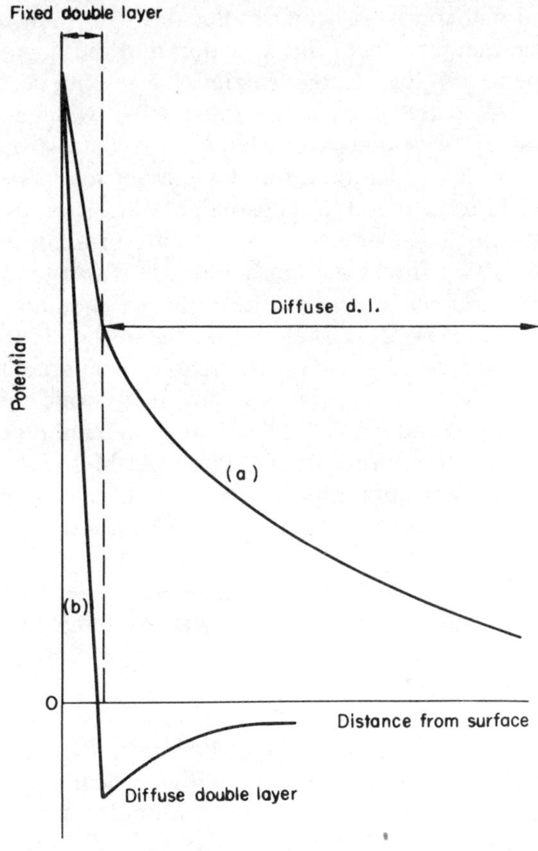


Figure E.2 Stern double layer

$+e_0$, $-e$ and $+(e - e_0)$. The two possibilities are illustrated in figure E.2.

The capacity of metal–solution interfaces can be calculated from electrocapillary curves, or from the growth of the potential during the passage of a very small charging current. The data show that the Helmholtz and Gouy models are both unsatisfactory, but that the Stern model can form the basis for an explanation of the main experimental features. The picture of a metal–solution interface that can be built up from this evidence is somewhat as follows. The layer in direct contact with the metal will consist of water molecules with their dipoles partly oriented to the metal surface; the direction and the degree of orientation will depend on the charge carried by the metal. Interspersed among these are likely to be some ions; these will be ions, usually anions, with a relatively low hydration energy or a strong adsorptive affinity for the metal, which therefore have a net tendency to exchange a water molecule for a metal atom as neighbour. There will also be an adsorption into this layer of any organic solute present, either because of specific adsorption forces or simply because it will tend (as at an air interface) to be squeezed out of the aqueous medium. The charge associated with this layer is said to reside at the 'inner Helmholtz plane'. Strongly hydrated ions will be unable to approach the surface so closely, and will be separated from it by one or two water molecules. They are adsorbed at the 'outer Helmholtz plane', and beyond this will be the diffuse double layer. At high concentrations of electrolyte the concentration of adsorbed ions increases and the diffuse double layer is compressed, so that its equivalent net charge is closer to the surface. The properties of the interface then approximate more closely to those of a Helmholtz double layer. At very low concentrations the diffuse part of the double layer becomes relatively more important.

*See also* Fr, Pa.

**Electric units**

In the electrostatic system of units the unit electric charge $Q$ is defined as that charge which, when placed 1 centimetre from an identical charge in a vacuum, repels it with a force of 1 dyne. When placed in an electric field, this unit charge will experience a force of $E$ dynes tending to move it in the direction of the field, and $E$ then measures the electric field strength in electrostatic units (e.s.u.). Similarly, the potential difference between two points will be 1 e.s.u. if 1 erg of work is gained or performed when unit charge is moved from one point to the other. The electrostatic unit of resistance $R$ follows from Ohm's law:

$$\mathrm{d}Q/\mathrm{d}t = I, \qquad R = E/I$$

An alternative system of electromagnetic units (e.m.u.) is based on Ampère's theorem, which defines current in terms of its magnetic effect.

In the International system of units (SI), now adopted, the basic electric unit is the ampere, A. This is defined as the current $I$ which, if maintained in two parallel conductors of infinite length and negligible circular cross-section, at a distance of 1 metre apart in a vacuum, would produce a force between the conductors equal to $2 \times 10^{-7}$ newton per metre of length. The other common SI electric units are: for quantity of electricity, the coulomb, $\mathrm{C} = \mathrm{A\,s}$; for electric potential, the volt, $\mathrm{V} = \mathrm{J\,A^{-1}\,s^{-1}}$; for resistance, the ohm, $\Omega = \mathrm{V\,A^{-1}}$; and for capacitance, the farad, $\mathrm{F} = \mathrm{C\,V^{-1}}$.

Conversion factors for these quantities are given in the following table:

| Unit | e.m.u. | e.s.u. | Abs./Int. |
|---|---|---|---|
| 1 A | $1 \times 10^{-1}$ | $3 \times 10^{9}$ | 0.999 835 |
| 1 C | $1 \times 10^{-1}$ | $3 \times 10^{9}$ | 0.999 835 |
| 1 V | $1 \times 10^{8}$ | 1/300 | 1.000 330 |
| 1 $\Omega$ | $1 \times 10^{9}$ | $1/(9 \times 10^{11})$ | 1.000 495 |
| 1 F | $1 \times 10^{-9}$ | $9 \times 10^{11}$ | 0.999 505 |

The last column of the table shows the factors by which the 'international unit' must be multiplied to convert it into the absolute unit now employed. The 'international units' were practical units adopted at an international conference in 1918, and much of the data in the literature are given in these units. They were replaced by the absolute units in 1946.

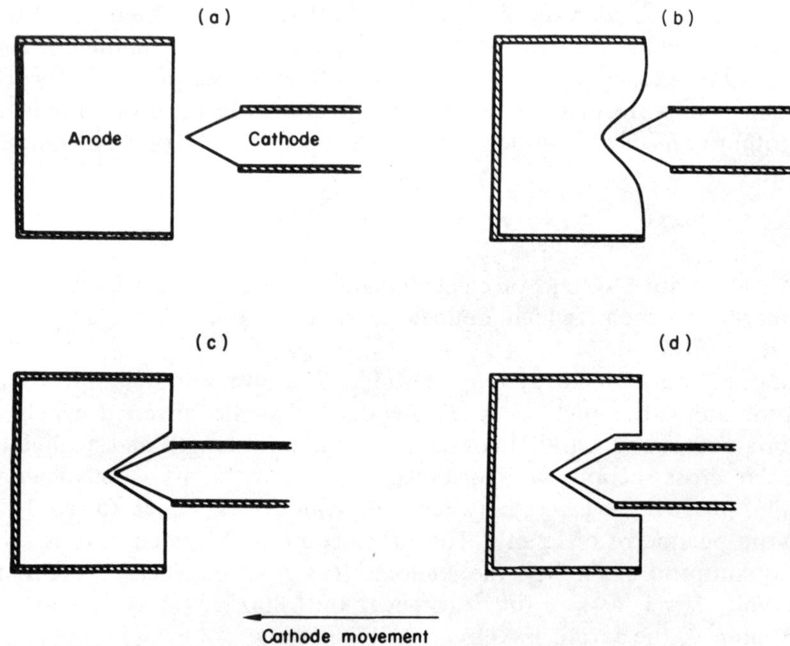


Figure E.3 Progress of an electrochemical machining operation

**Electrochemical equivalent**

The electrochemical equivalent of an element is the weight deposited or dissolved by the passage of 1 coulomb in an electrolytic process. By *Faraday's laws* (q.v.) it should equal the gram-equivalent weight divided by the *Faraday constant* (q.v.). Thus, the electrochemical equivalent of silver is 1.118 mg $C^{-1}$, and this multiplied by 96 490 (the Faraday constant) gives 107.88, the relative atomic mass of silver.

**Electrochemical machining**

In electrochemical machining the anodic dissolution of a metal is utilised in producing an article of the required shape or size. It has advantages over the normal mechanical processes for very tough metals, or where mechanical stresses are to be avoided.

Comparatively large amounts of metal are involved, and the electrochemical conditions used are similar in some respects to those of *electroforming* (q.v.). Very high current densities are used, and an electrolyte of high conductance must circulate through the gap between cathode and anode. Sodium chloride is often used. Hydrogen is then evolved at the cathode, and the metal dissolved at the anode is precipitated as its hydroxide and removed by a filtering unit in the circulating system. Efficient pumping is necessary to maintain a high rate of circulation and to remove the products of electrolysis and the heat generated from the neighbourhood of the electrodes.

A possible process is pictured in figure E.3. The sides of the anode and cathode are insulated, e.g. by a layer of plastic. The current density at the anode is highest opposite the tip of the cathode, and dissolution is most rapid here. The cathode is rigidly mounted, and constantly advanced at a rate to keep the gap constant, and to avoid sparking on the one hand, or a falling off of the current on the other. The required shape is eventually produced, being determined by the dimensions of the cathode and those of the (constant) gap between the electrodes.

*See also* P.

**Electrochemical series**

In the electrochemical series the metals are listed in the order of their chemical reactivity, the most active at the top and the least active at the bottom. In the broader sense it is not necessary to limit the series to metals, but may be carried through the electropositive (non-metallic) elements (table A.III, p. 244).

As applied to metals, the series was first established by laboratory experiments to discover which metal would displace others from their salts. Thus a strip of clean zinc immersed in a solution of copper sulphate is soon covered by a deposit of metallic copper while the zinc passes into solution as $Zn^{2+}$. This process is really an oxidation–reduction transfer of electrons:

$$\begin{array}{rl} & Zn \rightarrow Zn^{2+} + 2e \\ & Cu^{2+} + 2e \rightarrow Cu \\ \hline \text{i.e.} & Cu^{2+} + Zn \rightarrow Cu + Zn^{2+} \end{array}$$

In a similar way copper will displace silver from a solution of its salts.

To obtain a more accurate and reproducible series it is best to use the more exact *electrode potential* (q.v.), or oxidation–reduction potential. It is in this way that the original series applied to metals has been extended to non-metallic elements. Metals which will liberate hydrogen gas from dilute acids will be above hydrogen in the series, with a negative electrode potential, while those metals and non-metallic elements which will not liberate hydrogen from dilute acids will be below hydrogen in the series and will have positive electrode potentials.

**Electrode**

An electrode consists essentially of two conductors, one electronic and the other electrolytic, in contact. At the surface of separation (e.g. metal/metal ion in solution) a difference of potential, the electrode potential, exists. In principle, the work done in bringing a unit positive charge from infinity to the interface provides a measure of this potential; no such experiment is possible in practice.

The electrodes of a cell are the terminals or poles by which current enters or leaves the cell. The electrodes are in external metallic contact when the cell is working, and a stream of electrons flows through this contact from *anode* (q.v.) to cathode (metallic conduction). The same current is transported through the electrolyte of the cell by ions which give up or receive electrons at the electrode surfaces (electrolytic conduction), a process resulting in chemical reaction.

A *reversible galvanic cell* (q.v.) consists of an electrolyte and two reversible electrodes. The simplest type of reversible electrode or half-cell is an element in contact with a solution containing its ion. When a metal, M, is immersed in a solution of its ions, $M^+$, an electric double layer is formed, separation of charge occurs, and a potential difference is set up between M and the solution. In some metals (e.g. Zn) the atoms lose electrons, passing into solution as $M^+$:

$$M \rightarrow M^+ + e$$

a process which would result in the accumulation of liberated electrons in the metal, which would therefore become negatively charged with respect to the solution. On the other hand, with some metals (e.g. Cu) the reverse process occurs: $M^+ + e \rightarrow M$. This leads to a deficit of electrons in the metal, and the metal becomes positively charged with respect to the solution. In general, the number of ions deposited or released in this way is very small, and the tendencies for the two processes to occur are soon reduced to zero by the potential differences so established. As a negative potential develops on M, the rate of its ionisation decreases and, on the other hand, the rate at which ions are discharged increases, until the equilibrium

$$M \rightleftharpoons M^+ + e$$

is established when the two rates are equal; the final potential difference, the *electrode potential* (q.v.), which is established depends on the activity of the ions in solution and the temperature.

Simple elemental electrodes are not confined to metals in equilibrium with their ions; e.g. a *gas electrode* (q.v.) in which the gaseous element is in contact with its ions in solution at the surface of a platinum electrode is:

$$\tfrac{1}{2}Cl_2(g) + e \rightleftharpoons Cl^-$$

All electrode equilibria involve two opposing reactions, an oxidation and a reduction reaction:

$$\text{oxidised state } + ne \underset{\text{oxidation}}{\overset{\text{reduction}}{\rightleftharpoons}} \text{reduced state}$$

The electrode potential, $E$, of such an electrode is given by

$$E(\text{O, R}) = E^\ominus(\text{O, R}) + \frac{RT}{nF} \ln \frac{a(\text{oxidised state})}{a(\text{reduced state})}$$

where $E^\ominus(\text{O, R})$ is the standard electrode potential, i.e. the potential when $a$(oxidised state) = $a$(reduced state). When two such reversible electrode systems are connected without the free mixing of the solutions, a simple galvanic cell is formed and electron transfer will take place through the external circuit when the electrodes are connected externally. The separate electrode reactions are $M_1 \rightarrow M_1^+ + e$ and $M_2^+ + e \rightarrow M_2$, and the cell reaction is

$$M_1 + M_2^+ \rightarrow M_1^+ + M_2$$

in which $M_1$ is oxidised to $M_1^+$ and $M_2^+$ is reduced to $M_2$. The *Daniell cell* (q.v.) is a simple example of this type in which $M_1$ is zinc and $M_2$ is copper.

The potential difference between the electrode and the solution cannot be measured because there is no method of establishing a contact with the solution without setting up another electrode and, hence, another electrode potential. The *hydrogen electrode* (q.v.) is defined as a standard electrode of zero potential difference, so, using this as reference, the potential difference of any other electrode system can be measured.

In representing electrodes or half-cells diagrammatically, vertical lines are used to indicate the junction of two phases, e.g. $Zn^{2+} | Zn$.

Two main types of electrode are recognised: (1) indicating electrodes, the potential of which varies with the concentration (activity) of an ion in solution, and (2) reference electrodes, which are used to complete the cell system (i.e. comprising an indicating and a reference electrode). The potential of the reference electrode is not affected by the ion of the indicating electrode.

*Indicating electrodes*

(a) Electrodes of the first kind, i.e. an element in contact with a solution of its ions, e.g. $Zn \rightleftharpoons Zn^{2+} + 2e$, for which

$$E(Zn^{2+}, Zn) = E^{\ominus}(Zn^{2+}, Zn) + \frac{RT}{2F} \ln a(Zn^{2+})$$

Since the zinc is present as a metal in its standard state, $a(Zn) = 1$. The standard electrode potential, $E^{\ominus}(Zn^{2+}, Zn)$, is the potential when $a(Zn^{2+}) = 1$.

For the chlorine electrode $\frac{1}{2}Cl_2(g) + e \rightleftharpoons Cl^-$, in which electronic contact between the gas and solution is established by an inert platinum electrode:

$$E(Cl_2, Pt, Cl^-) = E^{\ominus}(Cl_2, Pt, Cl^-) + \frac{RT}{F} \ln \frac{p^{1/2}(Cl_2)}{a(Cl^-)}$$

In this instance the standard electrode potential $E^{\ominus}(Cl_2, Pt, Cl^-)$ is the electrode potential when the gas is present at 1 atmosphere pressure and $a(Cl^-) = 1$. Amalgam, gas, hydrogen and metal electrodes are in this class.

(b) Redox electrodes (see *redox electrode system*), e.g. $Fe^{3+}$, $Fe^{2+}$, Pt, in which a platinum electrode dips into a solution containing both the oxidised and reduced states, $Fe^{3+} + e \rightleftharpoons Fe^{2+}$, for which

$$E(Fe^{3+}, Fe^{2+}) = E^{\ominus}(Fe^{3+}, Fe^{2+}) + \frac{RT}{F} \ln \frac{a(Fe^{3+})}{a(Fe^{2+})}$$

(c) Cationic-responsive glass electrodes, e.g. the normal *glass electrode* (q.v.), which is reversible to $H^+$ in solution:

$$E_G = E_G^{\ominus} + \frac{RT}{F} \ln a(H^+)$$

(d) Other hydrogen-ion-indicating electrodes e.g., the *antimony–antimony oxide electrode* (q.v.), the *quinhydrone electrode* (q.v.).

(e) Cation-reversible electrodes, consisting of a metal, one of its insoluble salts, another insoluble salt of the same anion but a different cation dipping in a solution containing the common cation; e.g. the electrode $Ca^{2+} | CaC_2O_4(s) | PbC_2O_4(s) | Pb$ behaves as a reversible calcium electrode:

$$E(Ca^{2+}, Ca) = E' + \frac{RT}{2F} \ln a(Ca^{2+})$$

where $E'$ is not a true standard electrode potential but includes the solubility products of the sparingly soluble salts.

*Reference electrodes*

These electrodes, often known as electrodes of the second kind, generally consist of a metal in contact with a solution saturated with a sparingly soluble salt of the metal which also contains an additional salt with a common anion. Examples include $AgCl(s), Ag | KCl(aq)$; $Hg_2SO_4(s), Hg | K_2SO_4(aq)$ and $Hg_2Cl_2(s), Hg | KCl(aq)$. In all cases the electrode potential is governed by the activity of the anion in solution; i.e. the electrode is reversible with respect to the anion:

$$E = E' - \frac{RT}{nF} \ln a(A^{n-})$$

The *silver–silver chloride electrode* (q.v.) and *calomel electrode* (q.v.) are the most useful examples of this class.

*See also* De, I & J.

**Electrodeposition of metals**
Most metallic cations are readily discharged at potentials close to their equilibrium values, but with iron, nickel and other transition metals there is a significant *activation overpotential* (q.v.). The structure of the deposit produced depends greatly on the conditions of the electrolysis. When deposition takes place readily, and is unimpeded, relatively large crystals tend to form, which may continue the crystalline structure of the underlying metal. Good stirring, elevated temperature, and the use of a simple salt of the metal at a high concentration will favour this type of deposit. However, any constraint in the over-all process tends to lead to a fine-grained microcrystalline deposit. When this is the objective, it is common to use a complex salt of the metal as electrolyte, to add to the bath non-electrolytes which are strongly adsorbed at the surface, or to work at high current densities and overpotentials.

When the electrolyte contains more than one cation, their simultaneous discharge becomes possible. This will occur if their deposition potentials (under the conditions used) are close to one another, and it can lead to *alloy electrodeposition* (q.v.). The most general case, however, is of course where hydrogen ion is the second cation, and the simultaneous evolution of hydrogen is a common accompaniment of metal deposition. The amount of current that will be dissipated in this way can be calculated if the current–potential curves for the two processes are known. It must be remembered that the curve for $H_2$ evolution depends on the nature of the cathode surface as well as on the pH of the solution. Figure E.4 illustrates the deposition of zinc from a neutral solution of zinc sulphate. At the point at which the curves intersect the current efficiency for zinc deposition is 50%, but at high current densities it is much greater.

The mechanism of the discharge of a metal ion involves a number of stages, such as are shown in figure E.5. Initially the metal cation is in solution and associated with a complete hydration shell; in its final state it has attained a stable position in the metal lattice, but still has partial ionic character on account of the delocalised bonds associated with a metallic lattice. The intermediate stages are thought to involve separate steps in which the ion first becomes adsorbed on the surface as an 'ad-ion' and then progressively moves into positions in which it becomes more closely associated with the underlying metal structure,

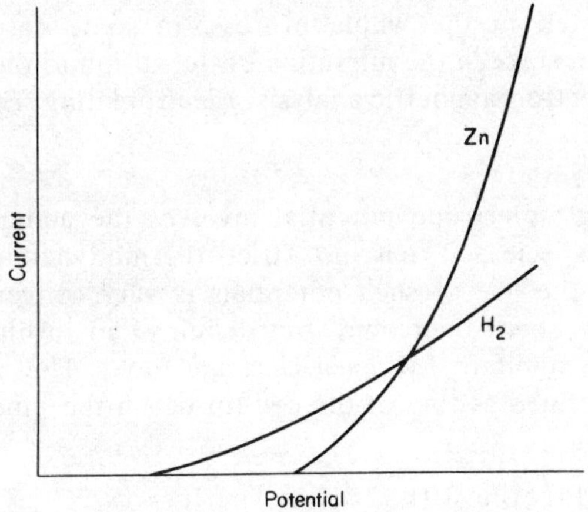


Figure E.4 Simultaneous discharge of zinc and hydrogen

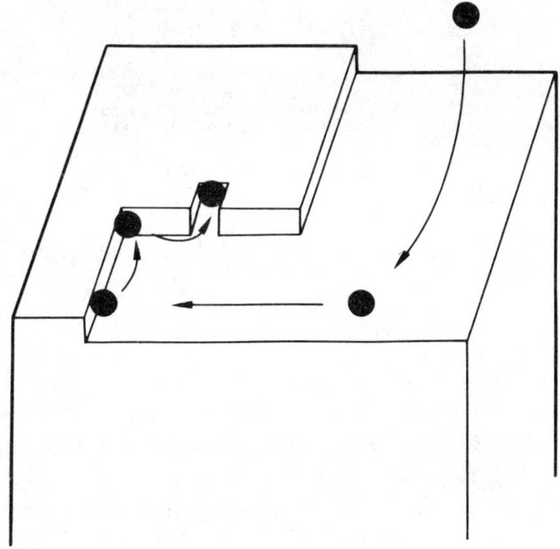

Figure E.5 Stages in the deposition of a metal ion

with a gradual loss of its water or ligand molecules. The rate-determining step in the whole process, in some cases at least, is thought to be a stage in the migration of the ad-ion to the growth site.

*See also* Electrogravimetric analysis; Electroplating; Electrorefining.

**Electrode potential**

Since the single electrode potential involves the activity of an individual ionic species, it has no strict thermodynamic significance. Nevertheless, the use of such potentials is very convenient, and the difficulty has been overcome by defining an arbitrary zero of potential—the standard *hydrogen electrode* (q.v.). The potential of an electrode is defined as that of the cell (in which the junction potential is ignored):

$$\text{Pt, H}_2(\text{g, 101 325 Nm}^{-2})\,|\text{H}^+(a=1)\vdots\ \text{X}^{n+}|\,\text{X}$$

$$E(\text{cell}) = E(\text{RH}) - E(\text{LH}) = E(\text{X}^{n+}, \text{X}) - E(\text{H}^+, \text{H}_2, \text{Pt}) = E(\text{X}^{n+}, \text{X})$$

implying a reaction of the type

$$\tfrac{1}{2}n\text{H}_2 + \text{X}^{n+} \rightarrow \text{X} + n\text{H}^+$$

for which

$$\Delta G = \Delta G^\ominus + RT \ln \frac{a(\text{X})a^n(\text{H}^+)}{a(\text{X}^{n+})p^{n/2}(\text{H}_2)}$$

Hence,

$$E(\text{X}^{n+}, \text{X}) = E^\ominus(\text{X}^{n+}, \text{X}) + \frac{RT}{nF} \ln \frac{a(\text{X}^{n+})}{a(\text{X})}$$

or, for an electrode reversible to an anion (e.g. chlorine),

$$E(\text{X}_2, \text{X}^{n-}) = E^\ominus(\text{X}_2, \text{X}^{n-}) - \frac{RT}{nF} \ln \frac{a(\text{X}^{n-})}{a^{1/2}(\text{X}_2)}$$

Thus, since $\Delta G^\ominus = -nFE^\ominus$, if $E^\ominus(\text{X}^{n+}, \text{X}) > 0$, $\Delta G^\ominus < 0$ and the forward reaction (reduction) is spontaneous ($\text{X}^{n+} + ne \rightarrow \text{X}$). If $E^\ominus(\text{X}^{n+}, \text{X}) < 0$, $\Delta G^\ominus > 0$, and oxidation is the spontaneous process.

The electrode potential thus depends on the standard electrode potential and the ratio of the activities of the oxidised and reduced states. Since for all electrodes (except redox) either the reduced or oxidised form of the electrode is in the standard state ($a = 1$), $E^\ominus(\text{X}^{n+}, \text{X})$ is the electrode potential when the activity of the ion to which the electrode is reversible is unity. Selected values of standard electrode potentials are listed in table A.III (p. 244).

If a cell is constructed with two metals dipping into solutions of their own ions, the metal which appears lower in the table is the positive pole and current flows inside the cell from the metal higher in the table to that lower. Thus metals with more negative potentials tend to displace from solution metals of more positive potential, e.g. in the *Daniell cell* (q.v.)

$$\ominus \quad Zn\,|Zn^{2+} \vdots\, Cu^{2+}|\,Cu \quad \oplus$$

$$E^{\ominus}(\text{cell}) = 0.339 - (-0.761) = 1.100\ \text{V}$$

and the cell reaction is $Zn + Cu^{2+} \rightarrow Cu + Zn^{2+}$. Metals with the greatest tendency to lose electrons have a high negative electrode potential; this is because the electrode at the right of the cell with a hydrogen electrode at the left becomes the more negative the more electropositive the metal.

Equilibrium constants can be calculated from standard electrode potentials. Thus, for the Daniell cell,

$$\log K = \log \frac{a(Zn^{2+})}{a(Cu^{2+})} = \frac{nFE^{\ominus}}{2.303RT} = \frac{2 \times 1.10}{0.059} = 37.28$$

Solubility products can be calculated in a similar way. Consider the cell

$$\ominus \quad Ag\,|Ag^{+}Cl^{-}|\,AgCl, Ag \quad \oplus$$

$$E^{\ominus} = 0.2225 - 0.799 = -0.5765\ \text{V}$$

for which the over-all reaction is

$$AgCl \rightarrow Ag^{+} + Cl^{-}$$

$$\log K_s = -0.5765/0.059;\ K_s = 1.7 \times 10^{-10}\ \text{mol}^2\ \text{dm}^{-6}$$

*Determination of standard electrode potentials*

(a) From the variation with the electrolyte concentration of the e.m.f. of a simple cell containing a standard hydrogen electrode, e.g.

$$\ominus \quad Pt, H_2(g, 101\,325\ \text{N m}^{-2})\,|H^{+}\ Cl^{-}|\,AgCl, Ag \quad \oplus$$

$$E(\text{cell}) = E^{\ominus}(\text{cell}) - \frac{2RT}{F}\ln a_{\pm} = E^{\ominus}(\text{cell}) - \frac{2RT}{F}\ln m_{\pm} - \frac{2RT}{F}\ln \gamma_{\pm}$$

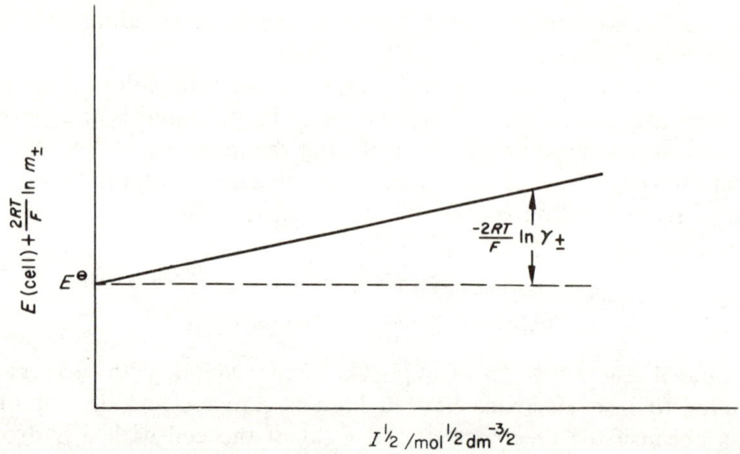


Figure E.6. Graph to determine $E^{\ominus}$(cell)

From the Debye–Hückel limiting equation $\log \gamma_{\pm} \propto I^{1/2}$; hence, the graph of $[E(\text{cell}) + (2RT/F) \ln m_{\pm}]$ against $I^{1/2}$ is linear (figure E.6); at $I = 0$, $\gamma_{\pm} = 1$; hence, the intercept is $E^{\ominus}(\text{cell}) = E^{\ominus}(\text{AgCl, Ag, Cl}^-)$. The deviation of the e.m.f. from this value gives the activity coefficient at various values of $m$.

(b) From the variation of the e.m.f. of galvanic cells with electrolyte concentration; this is useful for metals which form highly dissociated chlorides, e.g.

$$\ominus \quad \text{Zn} \left| \begin{matrix} \text{ZnCl}_2 \\ m \end{matrix} \right| \text{AgCl, Ag} \quad \oplus$$

for which

$$E(\text{cell}) = E^{\ominus}(\text{cell}) - \frac{RT}{2F} \ln 4m^3 - \frac{3RT}{2F} \ln \gamma_{\pm}$$

From this equation the graph of $E(\text{cell}) + (RT/2F) \ln 4m^3$ against $I^{1/2}$ is linear and of intercept $E^{\ominus}(\text{cell}) = E^{\ominus}(\text{AgCl, Ag, Cl}^-) - E^{\ominus}(\text{Zn}^{2+}, \text{Zn})$, from which $E^{\ominus}(\text{Zn}^{2+}, \text{Zn})$ can be obtained.

For bivalent metals which form highly dissociated soluble sulphates, similar cells using mercury–mercury(I) sulphate, or lead–lead sulphate electrodes may be used.

This method is also applicable to the determination of standard electrode potentials of halogen electrodes; thus, for the cell

$$\ominus \quad \text{Ag, AgCl}\,|\text{MCl}|\,\text{Cl}_2\text{, Pt} \quad \oplus$$

for which the cell reaction is $Ag(s)+\frac{1}{2}Cl_2(g)\rightarrow AgCl(s)$, and for which the e.m.f. is given by

$$E\text{ (cell)} = E^{\ominus}\text{(cell)} + \frac{RT}{2F}\ln p(Cl_2)$$

$E$(cell) is independent of the nature or concentration of the electrolyte, i.e. of M; thus, if $E$(cell) is determined at known gas pressures, $E^{\ominus}$(cell) and, hence, $E^{\ominus}(Cl_2, Cl^-)$ can be calculated.

(c) Indirect method from *equilibrium constant*‡ and *free energy*‡ data; the method is applicable when $E^{\ominus}$(cell) cannot be measured directly. Since

$$E^{\ominus}\text{(cell)} = -\frac{\Delta G^{\ominus}}{nF} = \frac{RT}{nF}\ln K_{\text{therm}}$$

the determination of $K_{\text{therm}}$ allows $E^{\ominus}$(cell) to be calculated for a hypothetical cell. Consider the determination of $E^{\ominus}(Sn^{2+}, Sn)$ for the hypothetical cell

$$\text{Sn}\,|Sn^{2+}\,\vdots\,Pb^{2+}|\,\text{Pb}$$

The over-all cell reaction is $Sn(s)+Pb^{2+}\rightarrow Pb(s)+Sn^{2+}$, and, hence,

$$K_{\text{therm}} = \frac{a(Sn^{2+})}{a(Pb^{2+})}$$

If a mixture of finely divided metallic tin and lead is shaken with a solution of their perchlorates and the equilibrium concentrations of tin and lead in solution are measured, $K_c$ can be calculated. If it is assumed that the activity coefficients of the two species are equal, $K_{\text{therm}}$ is given by the ratio of concentrations. Corrections can be applied for the activity coefficients using the Debye–Hückel activity equation—this correction is most necessary for standard redox potentials. For this system $K_{\text{therm}} = 2.98$ at 298 K. Therefore

$$E^{\ominus}\text{(cell)} = -0.014 = E^{\ominus}(Pb^{2+}, Pb) - E^{\ominus}(Sn^{2+}, Sn)$$

or $$E^{\ominus}(Sn^{2+}, Sn) = -0.014 - 0.126 = -0.140\text{ V}$$

(d) For alkali metals the difficulty in the determination of $E^{\ominus}(M^+, M)$

has been overcome by using an *amalgam electrode* (q.v.) which reacts only slowly with water.

*See also* De, G, I & J, Mi.

**Electrode reaction mechanisms**

As with ordinary chemical reactions, the main source of evidence concerning the mechanism of electrolytic processes comes from studies of their rates, which are conveniently measured by the current passing. But in place of temperature the main variable is now the potential at the electrode, and relations such as the *Tafel equation* (q.v.) provide the starting point for theoretical treatment. Complications such as those encountered in normal solution kinetics can be expected, however. The reaction will follow the easiest path, which will often comprise several consecutive steps, and these must be unravelled and the rate-determining step identified. Also, electrode processes are heterogeneous reactions, and the catalytic properties of the electrode and variations in the adsorption of substances present have to be taken into account.

The distinctive feature of an electrochemical reaction is the passage of an electron across the electrode–solution interface. Consider a cathodic process in which a univalent hydrated cation becomes discharged. Only the cations closest to the electrode will be involved—that is, those at the outer Helmholtz plane (see *electrical double layer*). Their discharge can be pictured as in the energy diagram (figure E.7). They must possess the energy necessary to reach the summit of the energy barrier, this extra energy being utilised in bond stretching and loosening the structure of the ion's hydration sheath. At the summit they will be in a state where electrons of suitable energy can be accepted, and the discharged ion will pass down into its stable position at the electrode surface. In the absence of an electric field the rate of the reaction could be written

$$v = kc_{\mathrm{A}} \exp(-E/RT) \qquad \text{(E.3)}$$

where $E$ is the chemical activation energy. With the ionic concentration $c_{\mathrm{A}}$ expressed in mol cm$^{-2}$ at the interface, $v$ is in mol cm$^{-2}$ s$^{-1}$, and multiplying by the charge per mole the current density is obtained as

$$j = Fv = Fkc_{\mathrm{A}} \exp(-E/RT)$$

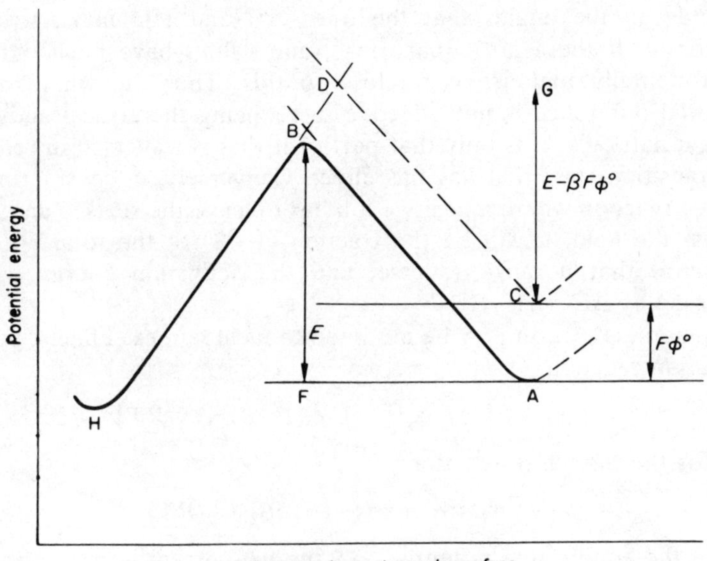


Figure E.7 Potential energy profile

This is an unreal situation, however. When the electrode system is set up, a potential at once arises at the interface, and increases until the two opposing processes $A + nH_2O \rightleftharpoons A^+(H_2O)_n + e$ are occurring at equal rates and equilibrium is reached. The equilibrium potential difference, $\phi°$, is now established, and an electrical contribution to the energy of the system must be taken into account. The left-hand side of the equation is relatively unaffected, consisting as it does of uncharged species, but the energy of the electron receives a contribution of $e\phi°$ or $F\phi°$ per mole, which must be added to the available energy of the right-hand system. With the approximation that during the transition through the double layer the energy varies linearly with distance, the line AB of the figure must be displaced upwards by an amount $F\phi°$, and becomes CD. The activation energy has thus been reduced from BF to CG. This amount is less than $F\phi°$, and the alteration in the activation energy is written $\Delta E = -\beta F\phi°$; the fraction $\beta$ is called the symmetry factor. A geometrical analysis of the diagram shows that $\beta$

depends on the angles that the lines AB and HD make with the horizontal. If these are equal, $\beta = \frac{1}{2}$, and values have been obtained experimentally that are very close to this. Thus the whole of the potential difference is not effective in reducing the critical activation energy; naturally, it is only that part of it that is traversed in reaching the transition state that has this effect. Conversely, in considering the reverse reaction where a positive ion has to leave the surface and move against the field, it will be the fraction $(1-\beta)$ of the total potential difference that it must traverse, and the activation energy will be increased by $\Delta E' = (1-\beta)F\phi°$.

Equation (E.3) can now be modified to include these effects. For the ionisation reaction

$$\vec{j} = Fk_1c_A \exp[-\{E+(1-\beta)F\phi°\}/RT] \qquad \text{(E.4)}$$

and for the discharge reaction

$$\overleftarrow{j} = Fk_2c_A \exp\{-(E-\beta F\phi°)/RT\} \qquad \text{(E.5)}$$

But at the equilibrium potential, $\phi°$, the *net* current is zero, i.e. $\vec{j} = \overleftarrow{j}$, and either measures the *exchange current density* (q.v.), $j_0$.

At any other electrode potential this equality will not apply, and a net current will flow. Suppose that the electrode potential is made more negative by an *overpotential* (q.v.) $\eta$. By the same argument as before, the activation energy for discharge of the cation is reduced by an amount $\beta F(\phi°+\eta)$, and that for ionisation increased by $(1-\beta)F(\phi°+\eta)$. In place of equations (E.4) and (E.5), therefore,

$$\vec{j} = Fk_1c_A \exp[-\{E+(1-\beta)F\phi°+(1-\beta)F\eta\}/RT] \qquad \text{(E.6)}$$

$$\overleftarrow{j} = Fk_2c_A \exp\{-(E-\beta F\phi°-\beta F\eta)/RT\} \qquad \text{(E.7)}$$

The expressions need not be kept in this extended form, however, since equations (E.4) and (E.5) represent the exchange current density $j_0$ at the standard equilibrium potential. Substituting this value in equations (E.6) and (E.7) gives the result of the applied overpotential as

$$\vec{j} = j_0 \exp\{-(1-\beta)F\eta/RT\}$$

and

$$\overleftarrow{j} = j_0 \exp(\beta F\eta/RT)$$

and for the net current

$$j = \overleftarrow{j} - \overrightarrow{j} = j_0 \exp(\beta F\eta/RT) - j_0 \exp\{-(1-\beta)F\eta/RT\}$$
$$= j_0 \exp[\beta F\eta/RT - \{-(1-\beta)F\eta/RT\}]. \quad \text{(E.8)}$$

At high cathodic overpotentials $\overrightarrow{j}$ becomes negligibly small, and

$$j = \overleftarrow{j} = j_0 \exp(\beta F\eta/RT)$$

or

$$\ln j = \ln j_0 + \beta F\eta/RT \quad \text{(E.9)}$$

This is equivalent to the Tafel equation, which thus provides experimental support for the theory outlined. A similar result can be written for the ionisation current obtained when the potential is much more positive than the equilibrium value.

When the overpotential is small ($\eta < 0.01$ V), neither of the opposing processes is negligible, and both must be taken into account. Equation (E.8) can be simplified at low $\eta$, however, by replacing each exponential by the first term in its expansion. Then, if $\beta = \frac{1}{2}$,

$$j = j_0(F\eta/2RT) - j_0(-F\eta/2RT) = j_0 F\eta/RT \quad \text{(E.10)}$$

At low overpotentials, therefore, the net current is proportional to the overpotential. This linear portion of the $j$–$\eta$ curve is illustrated in the entry *exchange current density* (figures E.12 and E.13), the slope of the curve in any particular case being governed by the magnitude of $j_0$ for the electrode process considered.

When these results for a single-electron transfer process, described by equation (E.9) at high overpotentials and by equation (E.10) at low overpotentials, are incorporated into expressions for the over-all mechanisms of electrolytic reactions, the simplifications used in arriving at them must be recognised. In particular, the effective area of the electrode was assumed to be constant, and the effects of adsorption upon it were neglected. Also, if $c_A$ and $\phi$ in the equations are to represent the ionic concentration of the bulk solution and the potential difference between this and the electrode, respectively, they must be corrected for the potential drop in the diffuse double layer, which has been ignored. Again, $\beta$ is a constant for the approximation that the energy–distance relations can be represented by straight lines; but the rounding off that actually occurs near the summit results in $\beta$ decreasing with increasing $\eta$ in cases where the activation energy is small and the overpotential is high.

The elucidation of the mechanism of an electrode reaction will depend in the first place on a knowledge of the chemistry of the reaction, the number of electrons involved and the constituents of the solution. Next, the Tafel slope (see *activation overpotential*) will give the value of $\alpha$, the transfer coefficient. This would be equal to $\beta$, the symmetry factor, for a simple electron transfer; but more generally it will be some multiple of $\beta$, depending on the number of electrons involved and other complicating features in the over-all process. The order of the reaction with respect to each constituent can be obtained by altering the concentrations one at a time and observing the effect on the current density at constant potential. Further help in identifying steps in the over-all process may come, as in ordinary kinetic studies, from isotope effects, alterations of pH and effects of additives. Finally, just as relaxation methods have been used in ordinary kinetics, the transient effects of current pulses, or periodically varying currents, can be analysed to identify intermediate steps and to provide information about adsorption behaviour at the electrode surface. Some illustrations of reasearch into mechanisms are given under *hydrogen evolution reaction* (q.v.).

*See also* B & R; and Thirsk, H. R. and Harrison, J. A. (1972), *Guide to the Study of Electrochemical Kinetics*, Academic Press.

**Electrodialysis**

Electrodialysis is an accelerated form of dialysis. The colloidal solution to be purified is contained within a membrane permeable to electrolyte ions, and two electrodes in the outside solution apply an electric field across the column of colloid. The ions to be removed move in each direction into the outer solution, and may be washed away with fresh solvent.

**Electroforming**

In the electroforming process a thick deposit of metal is electroplated on to a mould, which may be of metal or other material (such as plastic) which has been given a conducting coating. The deposit is then removed, and gives a perfect replica of the surface plated. The process is therefore a form of *electroplating* (q.v.) and involves no new principles. Uses include the making of electrotype by plating copper on to a plastic sheet bearing the required impression, and a somewhat similar process is used for making the master plates from which gramophone records are pressed.

**Electrogravimetric analysis**

The substance to be determined is deposited electrolytically on a suitable electrode, which is weighed before and after the experiment. Conditions, e.g. current density, may have to be chosen to give a pure, adherent deposit, and the solution is stirred to minimise *concentration overpotential* (q.v.).

The method is applicable to mixtures if correct conditions are chosen. Suppose, for example, that a copper–nickel solution is to be analysed. Copper deposits readily at a cathode potential close to the equilibrium value for copper, $E^\ominus(Cu^{2+}, Cu) = +0.34$ V; $E^\ominus$ for nickel is $-0.25$ V, but there is an appreciable *activation overpotential* (q.v.) for nickel deposition, so that a still more negative potential will be needed. An applied current will thus first deposit copper exclusively at a cathode potential of about $+0.3$ V. Towards the end of the deposition the concentration of copper ions will begin to fall steeply, and the cathode potential will have to become more negative to maintain deposition. However, a cathode potential of $-0.25$ V would correspond to a copper ion concentration given by

$$-0.25 = +0.34 + (0.059/2) \log [Cu^{2+}]$$

or $[Cu^{2+}] = 1 \times 10^{-20}$ mol dm$^{-3}$. The deposition of copper is therefore analytically complete long before there is any danger of codeposition. The cathode can be washed and weighed, and the nickel then plated out in a similar way. In practice, successful separations are considered to require a difference in the theoretical potentials of at least 0.2 V.

Complex formation in solution can greatly modify a metal-cation reversible potential (e.g. copper is deposited from a cuprammonium solution at a cathode potential of $-0.05$ V), so this effect can be used to increase the versatility of the method, and even to alter the order in

which two metals are deposited. Complexing may also improve the adherence of the deposit.

If necessary the method may be refined by using a *potentiostat* (q.v.) to maintain the cathode potential at the required calculated value, and by making additions to the solution to inhibit unwanted side reactions, such as simultaneous hydrogen evolution.

**Electrokinetic effects**

When an *electrical double layer* (q.v.) exists at an interface between a mobile phase and a stationary phase, a relative movement of the two can be induced by applying an electric field, and conversely, an induced relative movement of the two will give rise to a measurable potential difference. Four effects have been studied, and are related as follows:

| Known e.m.f. applied, rate of movement measured | Movement of | Movement induced, potential difference measured |
|---|---|---|
| *Electrophoresis* (q.v.) | charged particles through liquid | Sedimentation potential or *Dorn effect* (q.v.) |
| *Electro-osmosis* (q.v.) | liquid past stationary solid | *Streaming potential* (q.v.) |

These effects are classed as electrokinetic effects. The potential governing them is clearly that at the boundary between the stationary

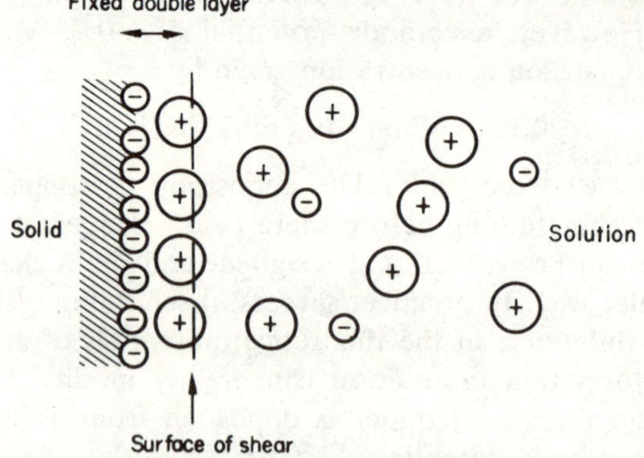


Figure E.8 Electrical double layer (schematic)

phase and the moving phase, and this is called the 'electrokinetic potential', or the zeta-potential, $\zeta$. Ions and molecules contained in the fixed double layer will be on one side of the surface of shear (figure E.8), while the diffuse double layer is contained in the other phase, so that the electrokinetic potential, measured by any of the above methods, is the potential difference at, or very close to, the outer Helmholtz plane. It can provide valuable information about the structure of the electrical double layer, and the adsorption characteristics of various materials.

**Electrolysis of brine**

Sodium chloride solutions are electrolysed on a very large scale for the production of sodium hydroxide and chlorine, and electrolysis is also used to produce hypochlorites and chlorates.

The deposition potential for sodium from an aqueous solution is −2.71 V, so that at a solid cathode hydrogen is always liberated in preference; one mole of hydroxide ion per Faraday is formed at the same time:

$$H_2O + e \rightarrow \tfrac{1}{2}H_2 + OH^-$$

If mercury is used as the cathode, the discharge of sodium ions will not produce a surface film of sodium metal, so the value −2.71 V is irrelevant. The sodium amalgamates with the mercury and diffuses away into the interior, so that the activity of sodium at the surface is extremely small. This, combined with the very high *hydrogen overpotential* (q.v.) at a mercury cathode, especially at fairly high current densities, ensures that sodium discharge is the predominant reaction.

At the anode the primary reaction is the discharge of chloride ions to give chlorine gas. Here, again, thermodynamic considerations would suggest that oxygen from the water ($E^\ominus = +0.81$ V at pH 7) would be discharged more readily than chlorine ($E^\ominus = +1.36$ V). It is not, partly because oxygen discharge would leave a highly acid film in contact with the anode, which would raise the theoretical $E$ by several tenths of a volt, and partly because of the high oxygen overpotential.

If the primary products are allowed to mix, hypochlorite is formed by the reaction

$$Cl_2 + 2OH^- \rightarrow Cl^- + ClO^- + H_2O$$

and if all the liberated chlorine is consumed in this way, 2 $F$ of electricity produce 1 mole of hypochlorite.

The hypochlorite may react further in two ways. In a warm acid solution, chlorate is formed by the reaction

$$2HClO + ClO^- \rightarrow ClO_3^- + 2H^+ + 2Cl^-$$

1 mole of chlorate resulting from the passage of 6 $F$. This reaction is very slow at low temperatures and in alkaline solutions, and the hypochlorite then undergoes a further reaction at the anode:

$$12ClO^- + 6H_2O \rightarrow 4ClO_3^- + 12H^+ + 8Cl^- + 3O_2 + 12e$$

Here 36 $F$ have produced 4 moles of chlorate (and 3 of oxygen), so that chlorate production is less efficient by this route.

These considerations form the basis for the following processes.

*Sodium hydroxide and chlorine production*

(a) Diaphragm cells. Steel cathodes and graphite anodes are used. Hydroxide ions could reach the cathode by diffusion or by electrolytic migration. To prevent the former, a diaphragm of treated asbestos surrounds the cathodes. Electrolytic transport is mainly by the chloride ions, which are present at a high concentration, but the hydroxide ion has a very high mobility, and to prevent it reaching the anode it is necessary to use counter-current circulation. The brine is fed continuously into the anode compartments, from which the chlorine is collected, and flows out through the cathode compartments; it is then concentrated and the remaining sodium chloride crystallised out. Hydrogen from the cathode compartments is also collected.

(b) Mercury cells. These are usually of (nearly) horizontal construction in which the mercury forming the cathode flows slowly along the floor of the cell in counter-current to the brine solution. Graphite anodes dip into the latter, and the chlorine liberated is led off. The amalgam concentration is not allowed to reach 0.2% sodium, as otherwise the fluidity of the mercury decreases, and also an appreciable amount of hydrogen would be discharged and would lower the current efficiency. The amalgam flows from the cell to a decomposer where it reacts with water:

$$Na(Hg) + H_2O \rightarrow Hg + NaOH + \tfrac{1}{2}H_2$$

Pure concentrated sodium hydroxide is produced directly by this method, but the voltage required is higher than in the diaphragm cell. Theoretically some of the extra electrical energy could be recovered from the spontaneous reaction in the decomposer, but practical difficulties have so far prevented this. The alkali-metal amalgams produced in mercury cells also have uses as reagents for various organic syntheses.

*Hypochlorite production*

Sodium chloride is electrolysed between electrodes usually of graphite. The temperature is kept low to minimise chlorate formation, and the electrodes are close together and arranged to ensure mixing. However, any hypochlorite reaching the cathode will be reduced, and in large plants it is now more common to produce NaOH and $Cl_2$ in diaphragm cells and form hypochlorite by mixing these.

Where sea-water is available, electrolysis provides an efficient method of sewage treatment. The sewage is mixed with sea-water and passed through an electrolytic cell. Hypochlorite is formed and rapidly sterilises the sewage, which then sediments out; the latter process is assisted by the gelatinous precipitate of magnesium hydroxide which forms in the alkaline solution.

*Chlorate production*

Sodium chloride is electrolysed between steel cathodes and graphite anodes. The temperature is kept at about 313 K (the graphite is oxidised at higher temperatures) and a pH of about 6 is maintained. A concentrated solution of chlorate is produced, which crystallises on cooling.

**Electrolysis of water**

Hydrogen and oxygen can be produced by a variety of methods, but where electricity is cheap the electrolysis of water has the advantage of producing gases of high purity. An electrolyte must be present to give a high conductivity; the high mobilities of the $H^+$ and $OH^-$ ions suggest the use of acids or alkali, but common materials withstand attack by alkalis much better than attack by acids, so sodium or potassium hydroxide, at a concentration giving the maximum conductance, is commonly used. Distilled water is added continuously to maintain

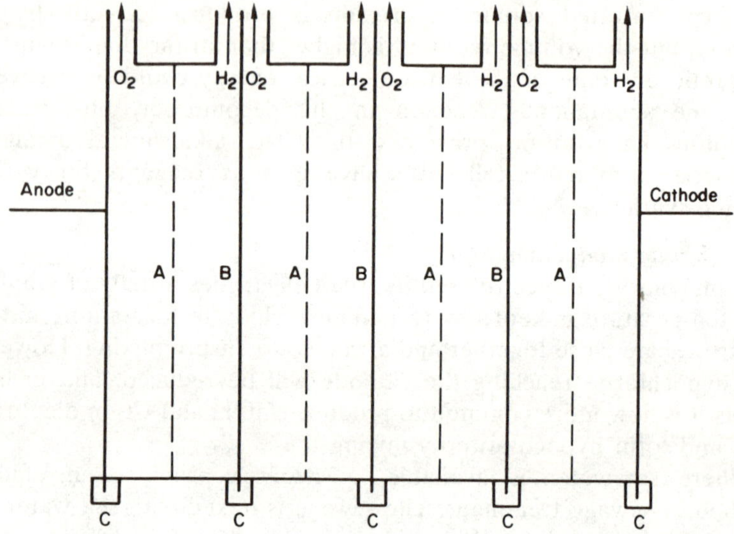


Figure E.9 Electrolysis of water in bipolar cell: A, diaphragms; B, bipolar electrodes; C, insulators

these conditions, and the operating temperature is about 343 K. The decomposition voltage of water is 1.23 V, but the *overvoltage* (q.v.) and resistance losses bring up the working voltage to about 2 V. The current efficiency can be as high as 99%, and the hydrogen produced is about 99.9% pure after drying.

The electrodes are normally of steel (the anode being nickel-coated to reduce the overpotential) and are separated by an asbestos membrane. A bipolar arrangement is often adopted, as shown diagrammatically in figure E.9, the intermediate plates acting as anodes on one side and as cathodes on the other.

*Heavy water production*

When water is electrolysed, the lightest isotope of hydrogen is preferentially liberated owing to differences in activation energy for the reactions at the cathode. There is therefore a gradual enrichment in deuterium of the liquid phase, and a practical measure of the extent of this effect is given by the separation factor, $S$, where

$$S = \frac{(H/D)_{gas}}{(H/D)_{liquid}} \quad (E.11)$$

Under the conditions used, the value of $S$ is about 6. A number of cells are arranged in a cascade so that the deuterium-enriched product of one cell feeds the next; almost pure $D_2O$ can be prepared in this way.

**Electrolytic cell**

The interconversion of electrical and chemical energy can be effected in a *primary cell* (q.v.), in which a spontaneous chemical process is used to produce electricity; or in the converse process of electrolysis, in which electrical energy from an outside source is used to bring about a chemical reaction.

An electrolytic cell normally consists of two electrodes dipping into an electrolyte solution, or of two solutions in contact through a porous diaphragm, with a suitable electrode in each. The electrode material will be chosen for its chemical properties and also, perhaps, for its catalytic effect upon the desired reaction at its surface. When such a cell is set up, equilibrium conditions are rapidly established; the electrodes acquire potentials at which there is no net tendency for electrons to move in either direction across the electrode–solution interface, and the corresponding *electrical double layer* (q.v.) is set up at each electrode surface. If now a gradually increasing e.m.f. is applied from an external source, electrolysis will begin when this reaches the *decomposition voltage* (q.v.). The availability of electrons at one electrode (the cathode) is now high enough to cause a reduction process, such as $e + Ag^+ \rightarrow Ag$ or $e + Fe^{3+} \rightarrow Fe^{2+}$, to proceed at a finite rate, while at the other electrode (anode) an oxidation process, such as $I^- \rightarrow \frac{1}{2}I_2 + e$ or $Ag \rightarrow Ag^+ + e$, supplies electrons to the electrode. The external battery can be likened to an electron pump, transferring electrons from anode to cathode. The corresponding current passing through the solution is carried by ions migrating to the electrodes.

A part of the electrical energy is expended in overcoming the resistance of the electrolyte. From Ohm's law this voltage drop is given by $\Delta E = IR$ for a current $I$ A passing through a resistance of $R\ \Omega$. The remainder of the applied e.m.f. must exceed $E^\ominus$, the equilibrium value of the cell reaction, for electrolysis to occur, and $E-E^{\ominus}$, the driving force of the reaction, is called the *overvoltage* (q.v.). The rate of the reaction is measured by the magnitude of the current passing, and this increases (within limits) with increasing overvoltage. The extent of chemical reaction is governed by *Faraday's laws* (q.v.).

*Examples*

(1) The current-carrying ions are discharged at the electrodes. In the electrolysis of moderately concentrated HCl solution, with inert platinum electrodes and a voltage of about 1.3 V, $H_2$ is evolved at the cathode and $Cl_2$ at the anode. In the solution the current is carried by the migration of $H_3O^+$ to the cathode and $Cl^-$ to the anode; in the outside circuit electrons flow from anode to cathode:

At cathode $\quad H_3O^+ + e \rightarrow \frac{1}{2}H_2 + H_2O$

At anode $\quad Cl^- \rightarrow \frac{1}{2}Cl_2 + e$

(2) A difficultly discharged anion leads to the decomposition of water at the anode to give $H^+$, $O_2$ and electrons, e.g. electrolysis of $CuSO_4$ solution with inert Pt electrodes. At the cathode Cu is deposited: $Cu^{2+} + 2e \rightarrow Cu$. At the anode $O_2$ is evolved and the solution adjacent to the electrode contains $H_2SO_4$. Sulphate ions migrate to the anode but are not the species discharged. Thus the anode reaction is

$$2H_2O \rightleftharpoons 2H^+ + 2OH^-$$
$$2OH^- \rightarrow H_2O + \tfrac{1}{2}O_2 + 2e$$

i.e. $$2H_2O \rightarrow 2H^+ + \tfrac{1}{2}O_2 + 2e$$

(3) A difficultly discharged cation leads to the decomposition of water at the cathode with the production of $OH^-$ and $H_2$, e.g. electrolysis of NaCl solution with Pt electrodes. The cathode reaction is represented

$$H_2O \rightleftharpoons H^+ + OH^-$$
$$H^+ + e \rightarrow \tfrac{1}{2}H_2$$

i.e $$H_2O + e \rightarrow OH^- + \tfrac{1}{2}H_2$$

(4) Reacting electrodes, e.g. the electrolysis of $CuSO_4$ solution between Cu electrodes. $Cu^{2+}$ and $SO_4^{2-}$ are the current-carrying ions. Cu is deposited on the cathode and Cu passes from the anode into

solution, there being no change in the electrolyte concentration. Cells of this type are used in the *electrorefining* (q.v.) of metals and as a *coulometer* (q.v.).

**Electrolytic oxidation of organic compounds**

As a synthetic method, electrolytic oxidation has probably proved, up till now, less useful than electrolytic reduction. An aliphatic hydrocarbon can be oxidised through the following stages:

$$\text{Hydrocarbon} \rightarrow \text{Alcohol} \rightarrow \text{Aldehyde} \rightarrow \text{Acid} \rightarrow CO_2$$

but obtaining the desired product in good yield necessitates close control to maintain the best conditions, and well-tried chemical methods often provide a more satisfactory alternative.

As with electrolytic reductions, the information that has accumulated in the past is now being supplemented by more searching studies of mechanism. The primary step in an oxidation may be the direct transfer of an electron, to yield a carbonium ion or other free radical which then reacts with other species in the adsorption layer; or it may be the direct oxidation of a 'carrier', such as the manganese(II) ion or some other substance that readily enters into a reversible oxidation–reduction process. Also, the solvent may be involved, and in aqueous media the $OH^-$ ion may lose an electron to become an OH radical, which may then give rise to hydrogen peroxide or to an adsorbed oxygen atom, or may react directly with the organic substrate. Finally, at a lead dioxide electrode, which is often used as anode, the electrode surface itself may be involved in the oxidation.

An anodic oxidation with a long history is the Kolbe synthesis of hydrocarbons, exemplified by the electrolysis of an alkali-metal acetate solution to give ethane:

$$2CH_3.CO_2^- \rightarrow 2CO_2 + C_2H_6 + 2e$$

This is a general reaction, and a mixture of acids can be used to obtain a hydrocarbon derived from both; the method does not apply to aromatic acids or dicarboxylic acids, and substituents have a large effect on the yield. The reaction takes place in both aqueous and non-aqueous solutions, and can be explained by the scheme

$$R.CO_2^- \xrightarrow{-e} R.CO_2^{\cdot} \longrightarrow CO_2 + R^{\cdot}\ ; \quad 2R^{\cdot} \longrightarrow R_2$$

However, some of the features of the reaction in aqueous solution seemed to point to the intervention of hydroxyl radicals or hydrogen peroxide by a mechanism such as

$$OH^- \rightarrow OH^\cdot + e; \quad 2OH^\cdot \rightarrow H_2O_2$$

$$2R.CO_2^- + H_2O_2 \rightarrow 2R.CO_2^\cdot + 2OH^-; \quad 2R.CO_2^\cdot \rightarrow R_2 + 2CO_2$$

This is the 'hydrogen peroxide theory'.

The Kolbe reaction occurs at a high anode potential. At a lower value on a platinum anode acetic acid can be oxidised to glycolic acid, glyoxylic acid and finally oxalic acid:

$$CH_3CO_2H \rightarrow CH_2OH.CO_2H \rightarrow CHO.CO_2H \rightarrow HCO_2.CO_2H$$

Characteristic of aromatic compounds is the anodic oxidation of benzene, which can yield the following products:

Benzene → Phenol → Catechol; Phenol → Hydroquinone → Quinone → $CO_2H.CH{=}CH.CO_2H$ (Maleic acid)

Halogenations, and other substitution reactions, can also be carried out at an anode. On the industrial scale the method has been used for the preparation of iodoform from ethanol and KI. The process occurs at the iodine–iodide potential, so the primary step is presumably the discharge of iodide ion. The hypoiodite ion has been shown to participate in the subsequent reaction, so the process is conducted at a controlled, slightly alkaline, pH; pure iodoform precipitates from the aqueous solution in good yield. Anodic halogenation has also been found very convenient in fluorinations, since it avoids the use of elementary fluorine.

*See also* Al, L; and *Specialist Periodical Reports, Electrochemistry*, **2** (1972) (The Chemical Society, London).

**Electrolytic polishing**

The electrolytic polishing process is a selective electrolytic dissolution of metals in which projections in the surface are attacked more rapidly than depressions. The object may be to produce a bright surface, or one suitable for micrographic examination. The surface to be treated is made the anode, and a great variety of electrolytes have been used; perchloric, chromic and nitric acids are common constituents, but salt solutions or fused salts are more suitable in some cases.

The technique has not been fully explained, but the conditions under which electropolishing is most effective appear to be those in which the anodic dissolution of the metal is diffusion-controlled. As the metal dissolves, there is a layer of solution of high concentration at the anode, and this may give rise to a film of insoluble salt or oxide at the surface which will further hinder the diffusion of ions into the bulk of the solution. These factors reinforce the normal tendency for diffusion to take place more readily from projections than from concave areas of the surface, and thus give rise to the polishing effect.

**Electrolytic reduction of organic compounds**

Some organic compounds can be reduced at a cathode under reversible conditions: the quinhydrone electrode is a well-known example. In general, however, organic compounds are liable to be reduced at an inert cathode without giving rise to any definite potential; a variety of products may be obtained, depending on the conditions, and the reactions are usually complex and involve a number of consecutive or simultaneous steps. A large amount of empirical information has been accumulated in the past about individual reactions, and a modern resurgence of interest in the subject is centred on the use of the techniques now available to elucidate the nature of the successive stages in the over-all reaction.

The primary step may be any of the following: the direct transfer of electrons from the electrode to the organic compound; the formation of solvated electrons; or the discharge of hydrogen atoms at the electrode surface. Which of these occurs, and what ensues, will obviously depend on a number of factors. The nature of the solvent is important; aqueous solutions, mixed solvents and anhydrous solvents have all been used, and besides influencing the primary step the solvent may exert an effect by altering the composition of the adsorbed layer on the electrode surface, and by entering into chemical reaction with an unstable intermediate product in the chain of reaction steps. The electrode may exert an influence through its catalytic properties, and through its adsorption characteristics with respect to the various species present. Its hydrogen overpotential is also very important; when the compound under study is difficultly reducible, hydrogen will be evolved at a low-overpotential metal, and only an electrode of high overpotential will supply the energy necessary for the reduction to take place. Other factors that can affect the product are the concentration of the organic compound, the current density, the rate of stirring, the temperature and the presence in the electrolyte of acid, alkali or other catalytic agents, such as titanium or cerium salts.

The electrolytic reduction of aromatic nitro-compounds has been studied extensively. Nitrobenzene can be reduced to aniline in three stages, through nitrosobenzene and phenylhydroxylamine:

$$C_6H_5NO_2 + 2e + 2H^+ \rightarrow C_6H_5NO + H_2O$$
$$C_6H_5NO + 2e + 2H^+ \rightarrow C_6H_5NHOH$$
$$C_6H_5NHOH + 2e + 2H^+ \rightarrow C_6H_5NH_2 + H_2O$$

The first two steps are believed to proceed by direct electron transfer:

$$C_6H_5NO_2 + 2e \rightarrow C_6H_5NO_2^{2-} \xrightarrow{2H^+} C_6H_5NO + H_2O$$

and

$$C_6H_5NO + 2e \rightarrow C_6H_5NO^{2-} \xrightarrow{2H^+} C_6H_5NHOH$$

Nitrosobenzene is more easily reducible than nitrobenzene, and so does not accumulate in the solution, but it can be precipitated as it is formed if the nitrobenzene is reduced in the presence of hydroxylamine and $\alpha$-naphthylamine, which react with it to form an azo dye.

The final stage, to give aniline, only occurs rapidly at high-overpotential metals, and is probably a reduction by atomic hydrogen. At lower potentials the phenylhydroxylamine can react, according to conditions, in a variety of side reactions:

$$C_6H_5NHOH + C_6H_5NO \rightarrow H_2O + C_6H_5\overset{\overset{O}{\uparrow}}{N}{:}NC_6H_5 \qquad \text{Azoxybenzene}$$

$$2C_6H_5NHOH \rightarrow 2H_2O + C_6H_5N{:}NC_6H_5 \qquad \text{Azobenzene}$$

$$C_6H_5NHOH \rightarrow p\text{-}HO.C_6H_4NH_2 \qquad p\text{-Aminophenol}$$

The first stage in the reduction of a carboxylic acid group is to the corresponding aldehyde:

$$R.C\begin{matrix}{}^{\diagup\!\!\!\diagup O}\\ {}_{\diagdown OH}\end{matrix} + 2e + 2H^+ \longrightarrow R.C\begin{matrix}{}^{\diagup OH}\\ -H\\ {}_{\diagdown OH}\end{matrix} \longrightarrow R.C\begin{matrix}{}^{\diagup\!\!\!\diagup O}\\ {}_{\diagdown H}\end{matrix} + H_2O$$

Aldehydes and ketones may be further reduced:

$$R_2CO + 2e + 2H^+ \rightarrow R_2CHOH \xrightarrow{2e+2H^+} R_2CH_2 + H_2O$$

but a further possibility, at higher cathode potentials, is the formation of a free radical:

$$R_2CO + e + H^+ \rightarrow R_2C(OH)$$

and this is normally followed by dimerisation to the pinacol. Another well-known hydrodimerisation reaction is the industrial preparation of adiponitrile from acrylonitrile:

$$2CH_2{=}CH.CN + 2e + 2H^+ \longrightarrow \begin{matrix} CH_2.CH_2.CN \\ | \\ CH_2.CH_2.CN \end{matrix}$$

A different type of reactive intermediate that may occur, especially at lead and mercury electrodes, is an organometallic compound.

Further fields in which electrolytic reduction has been exploited are the hydrogenation of unsaturated compounds, the removal of aliphatic halogen atoms (e.g. $CCl_4 \rightarrow CHCl_3 \rightarrow CH_2Cl_2$) and the desulphonation of aromatic compounds. A wide variety of organic syntheses are possible by the electrolytic production of a reactive species in a suitable solution of the necessary reactants.

*See also* Al, L; and *Specialist Periodical Reports, Electrochemistry*, **2** (1972) (The Chemical Society, London).

**Electrometric titrations**

A titration is performed to determine the chemical equivalence of one reagent for another in a well-characterised reaction by observing the change in some property of the solution. The equilibrium of the reaction must lie far towards completion, so that the equivalence point is accompanied by a large and sudden change in the concentration of one of the reactants. When the reaction has proceeded sufficiently far to actuate the means of observation, the 'end-point' is reached. Chemical indicators which respond to change of pH or oxidation potential are well established. To overcome some of the limitations of visual indicators, e.g. in highly coloured solutions, instrumental methods have been developed to locate the end-point. *Amperometric titration* (q.v.), *conductimetric titration* (q.v.) and *potentiometric titration* (q.v.) are extensively used. They extend the advantages of titrimetric analysis to titrations not otherwise feasible because of their equilibrium constants.

**Electromotive force**

The e.m.f. (dimensions: $\mu^{1/2} m^{1/2} l^{3/2} t^{-2}$; units: $V = kg\ m^2\ s^{-3}\ A^{-1} = JA^{-1}\ s^{-1} = WA^{-1}$) between two points is that force which causes a current to flow between the points. The unit of electric potential, the volt, is the difference of potential between two points of a conducting wire carrying a current of 1 ampere, when the power dissipated between these points is 1 watt. The sign of the e.m.f. is defined so that a positive charge will tend to flow from a higher to a lower potential. Applied to a cell, the e.m.f. is the electric potential between two pieces of metal with identical composition, the ends of the chain of conducting phases. Thus in the *Daniell cell* (q.v.)

$$\ominus \quad Cu' \,|\, Zn \,|\, Zn^{2+} \,\vdots\, Cu^{2+} \,|\, Cu \quad \oplus$$

the e.m.f. is the difference between the potentials of the phases Cu' (i.e. the connecting wire) and the copper electrode; it is not the potential between the zinc and the copper electrodes. Any number of contact potentials in the external circuit always reduce to the Cu | Zn contact potential.

An e.m.f. measurement is only useful for a thermodynamically reversible cell, when $E$ is related to the free energy change of the chemical reaction by

$$\Delta G = -nFE \tag{E.12}$$

*Measurement of e.m.f.*
*See* Potentiometer.

**Electromotive series**
The electromotive series is a classification of the elements according to the values of the standard electrode potential (see *electrode potential*).

*See also* Table A.III, p. 244.

**Electro-osmosis**
Electro-osmosis is one of the *electrokinetic effects* (q.v.). If a potential difference is applied between the ends of a capillary tube containing electrolyte, or across a plug of finely divided material (which can be regarded as a bundle of capillaries), a movement of the liquid is observed. This is the reverse effect of *electrophoresis* (q.v.), where particles move through a liquid which is stationary. The effect can be studied in an apparatus such as that sketched in figure E.10. A plug of the finely divided material is in the centre of the tube, which is completely filled with liquid, and a potential of, say, 200 V is applied between the two calomel electrodes. An air-bubble trapped in the capillary measures the rate of movement of the liquid.

The effect is dependent on the *electrical double layer* (q.v.) at the interface, and if a plane surface can be assumed (i.e. if the curvature is negligible compared with the thickness of the diffuse double layer), the interface can be treated as a parallel plate capacitor. For steady conditions, the electrical force applied must balance the frictional force. Now the viscosity of the liquid $\eta$ is the force per unit area per unit velocity gradient. The velocity of the liquid is zero at the surface of shear and the velocity gradient can be written as $v/\kappa^{-1}$, where $\kappa^{-1}$ is

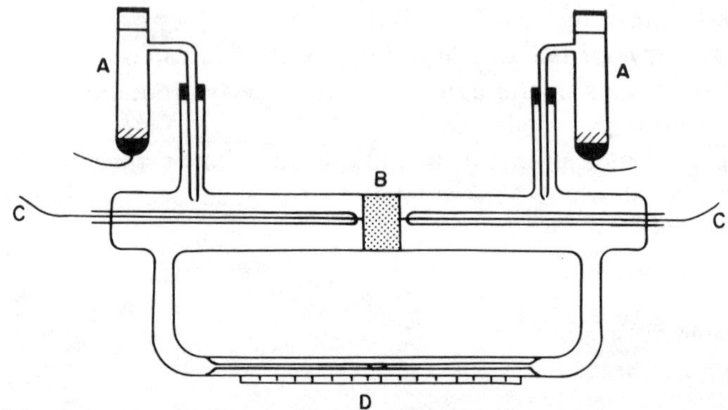


Figure E.10 Electro-osmosis cell: A, A, calomel electrodes; B, plug of material; C, C, leads to platinum gauze electrodes; D, air bubble and scale

the thickness of the diffuse double layer and $v$ is the velocity. If the net charge of the double layer is $\sigma$ C m$^{-2}$, the electric force for a potential difference $X$ is $X\sigma$. Equating this to the frictional force, $\eta v/\kappa^{-1}$ C m$^{-2}$, gives

$$u = v/X = \sigma\kappa^{-1}/\eta \tag{E.13}$$

for the velocity under unit potential gradient. If the formula for a parallel plate capacitor is now introduced,

$$\zeta = \frac{4\pi\kappa^{-1}\sigma}{\varepsilon}$$

where $\zeta$ is the electrokinetic potential across the interface. Combination with equation (E.13) gives

$$u = \zeta\varepsilon/4\pi\eta$$

For a plug of material of length $l$ and total effective cross-section $A$ across which a potential difference $E$ is applied, the volume of liquid transported in unit time becomes

$$V = A\zeta\varepsilon E/4\pi\eta l \tag{E.14}$$

$A$ cannot be accurately measured, but if $I$, the current passing during the experiment, is measured, and the conductivity $\kappa$ of the liquid is determined by means of the two Pt gauze electrodes (which also hold the plug in place), then

$$I = E/\text{resistance} = E\kappa A/l$$

and eliminating $A$ and $l$ from equation (E.14) gives

$$V = \zeta\varepsilon I/4\pi\eta\kappa$$

from which $\zeta$ can be found.

**Electrophoresis**

Electrophoresis is the movement, in an electric field, of charged particles through a liquid. The phenomenon has been studied in various ways. Tiselius (1937) perfected apparatus similar to that used in the *moving boundary method* (q.v.) of determining transport numbers, in which the boundary of a colloidal solution moving towards cathode or anode was observed by a Schlieren technique. Important pioneering work on the identification and separation of proteins was carried out with this. For very small quantities of material, the whole experiment can be performed on the stage of a microscope or ultramicroscope, using micro-electrodes and measuring the actual movement of a single particle across a calibrated scale in the eyepiece (figure E.11). Electrophoresis can also be carried out on strips of filter paper or other suitable material as a modification to chromatographic methods of analysis and separation.

Consider a small spherical particle of radius $r$ moving through a viscous medium; Stokes's law gives the steady velocity $v$ as

$$v = \text{driving force}/6\pi\eta r$$

The electric force driving the particle is $QX$, the product of the charge and the electric field, so for unit potential gradient the mobility of the

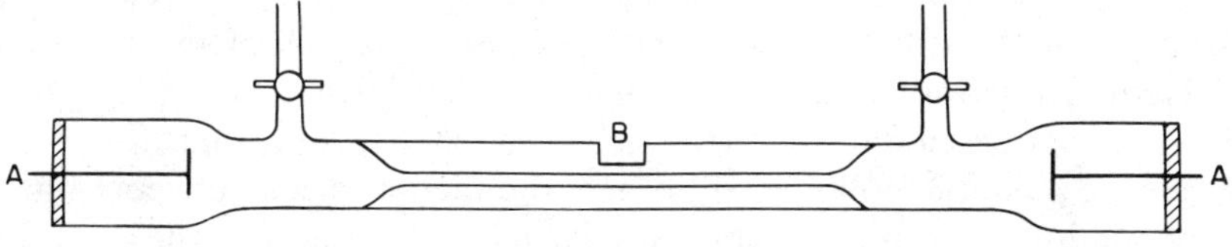


Figure E.11 Micro-cell for electrophoresis: A, A, electrodes; B, point of observation

particle is

$$u = Q/6\pi\eta r \qquad \text{(E.15)}$$

The equation is not useful if $Q$ and $r$ are both unknown. However, the potential $\psi$ at the surface of a sphere of charge $Q$ and radius $r$ is given by

$$\psi = Q/\varepsilon r$$

so by substituting in equation (E.15)

$$\psi = 6\pi\eta u/\varepsilon \qquad \text{(E.16)}$$

This is not immediately applicable to electrophoresis, however. The solid particle with its fixed double layer (net charge $Q$) is moving relative to a solution in which the diffuse double layer is distributed (see *electrical double layer*). The latter is equivalent to a charge $-Q$ spread out on a concentric sphere of radius $\kappa^{-1}$, where this is the thickness of the ionic atmosphere. The presence of this atmosphere reduces the mobility, and the potential at the surface of the particle, by the factor $1/(1+\kappa r)$ so that in place of equation (E.16) the zeta-potential (see *electrokinetic effects*) is given by

$$\zeta = \frac{6\pi\eta u}{\varepsilon(1+\kappa r)}$$

In the article on *electro-osmosis* (q.v.) a similar formula, but with 4 in place of the factor 6, is derived. Since electrophoresis is the reverse of electro-osmosis, the same expression should apply in both cases to the potential at the surface of shear between the two phases. The explanation of the apparent discrepancy is that instead of applying Stokes's law to a small sphere, the derivation of the electro-osmotic effect is based on the model of a parallel plate capacitor, i.e. on a large solid surface whose radius of curvature is negligible (compared with the thickness of the diffuse double layer). Closer analysis of the problem by Henry and Booth has shown that 4 is the correct factor for large particles, independent of their size and shape, but that for most systems, e.g. stable colloidal solutions, the factor varies between 4 and 6, depending on the size of the particle and the thickness of its atmosphere.

**Electroplating**

Electroplating consists in the *electrodeposition of metals* (q.v.), usually in a layer $10^{-2}-10^{-1}$ mm thick, for decoration or protection. The article to be plated need not necessarily be metallic; if not, a conducting layer must be formed on it before plating, e.g. by applying a graphite or metal powder, or treating with silver nitrate and reducing solution, or direct metallising in a vacuum.

Drastic cleaning of the article is essential, and may include degreasing in an organic solvent, treatment in alkaline detergent solution, electrolysis for a brief period at a high current density, and finally an acid dip. Rinsing is needed between each stage, and in large-scale practice the article may move automatically through the necessary series of tanks.

The composition of the electrolyte, and the current density and other conditions, must be adjusted to give a uniform and perfectly adherent deposit of very fine crystalline grain. Adherence is sometimes improved by first depositing a very thin layer of a metal which forms solid solutions with both the underlying metal and the metal to be subsequently deposited; a micro-crystalline deposit is usually favoured by the use of a complex salt as electrolyte, complex cyanides being used in most cases. Plating baths may also include buffering agents, small additions of surface-active agents found empirically to have a good effect on the structure of the deposit, and inert electrolytes. A bath should have good 'throwing power'; i.e. it should give a deposit of uniform thickness even when the article has projections (which are closer to the anode) or recesses (where the current density is likely to be reduced). Inert electrolytes have an influence here in determining the proportion of material brought to the surface by conduction; and the variation of *overpotential* (q.v.) with current density, and the diffusion rates and chemical stability of the various complex ions present in the surface layer, are all factors in determining throwing power.

Renewal of the electrolyte concentration at the cathode surface may be left to convective mixing and to the stirring effect of any hydrogen evolved, but additional stirring is usually provided either mechanically or by streams of air pumped into the bath. The anode is usually made of the metal being plated, and if the current efficiency is the same for cathode deposition and anode dissolution, the composition of the electrolyte will be maintained. The anode must not be allowed to become passive, as, for instance, a nickel anode tends to do at high current density, and additions to the bath or to the anode itself may be made to prevent this. The dissolution of the anode tends to produce an 'anode sludge', consisting of insoluble particles of the metal and impurities. These must not be allowed to contaminate the cathode surface, and so the anode is contained in a porous bag, or the bath liquid is continuously filtered. The composition of the bath should be kept under constant control, since any evolution of hydrogen at the cathode, or oxygen at the anode, will alter the pH. Any variations in current efficiency may alter the concentration of the cation being discharged, and a low concentration of an additive may gradually disappear through side reactions or by inclusion in the electrodeposit.

*See also* Lowenheim, F. A. (1963), *Modern Electroplating*, Wiley; Ollard, E. A. and Smith, E. B. (1964), *Handbook of Industrial Electroplating*, Iliffe.

**Electrorefining**

If two electrodes of a metal, such as copper, are immersed in a solution of one of its salts, and a small potential difference is applied between them, metal will dissolve from the anode, the more positive electrode, and deposit in corresponding amount on the cathode. The net chemical reaction is nil, and energy is required only to overcome the resistance of the solution and the (usually very small) *overpotential* (q.v.) at the electrodes.

The principle is very widely used in electrorefining. The impure metal is used as anode, and pure metal is obtained at the cathode. Any impurity metals with a more positive electrode potential than that of the metal being refined will remain undissolved at the anode, and may be recovered as 'anode sludge'. Any metal with an $E^{\ominus}$ more negative than that of the metal being refined will dissolve with it, but the cathode potential will be too low for its discharge and it will accumulate in the bath, and must eventually (or continuously in a circulating system) be removed for chemical treatment.

The conditions for a successful process include: (a) the conductance of the electrolyte should be high, to reduce the *IR* voltage drop; (b) the electrolyte should be selected with two further objectives in mind—it should be such as to give an adherent deposit, and it should give satisfactory dissolution of the anode (preventing any danger of this electrode becoming passive); (c) there must be no danger of a constituent of the anode sludge being taken up into solution by the electrolyte, as it would then be discharged at the cathode; (d) the current density should preferably be high, but must not approach the *limiting current density* (q.v.), or lead to the discharge of hydrogen.

The method is very extensively used for obtaining pure copper, for the refining of silver and gold, and as the final stage in the purification of many other metals.

**Exchange current density**

When the *activation overpotential* (q.v.) of a working electrode is plotted against the logarithm of the current density, a straight line is obtained at moderately high overpotentials (figure E.12). At low overpotentials there are departures from this relationship, for the log $j$ values must curve away to the value $j = 0$, $\log j = -\infty$ at $\eta = 0$. At this point, i.e. at the equilibrium potential, the two opposing processes of

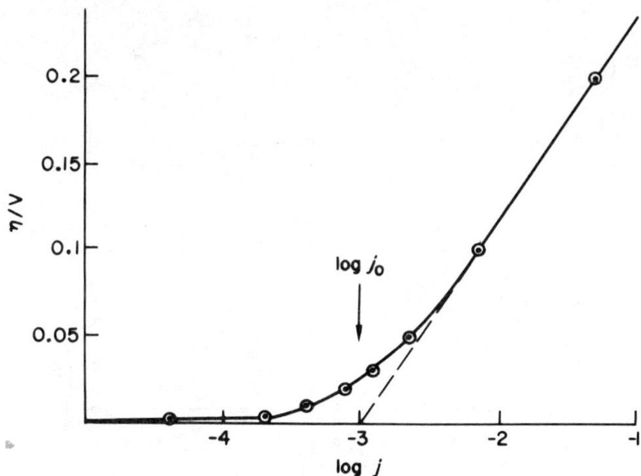


Figure E.12 Tafel diagram for an electrode process at 25 °C, with $j_0 = 1 \times 10^{-3}$ A cm$^{-2}$ and $\alpha = 0.5$

discharge and ionisation are in balance, and are proceeding at equal rates; if the cathodic and anodic current densities at the electrode are denoted $\vec{j}$ and $\overleftarrow{j}$, then $\vec{j}=\overleftarrow{j}$ at the reversible potential and the net current $j=\vec{j}-\overleftarrow{j}=0$. If the electrode is made slightly more negative, $\vec{j}$ will increase; $\overleftarrow{j}$ will be retarded but will still have a finite value, and the net current is the amount by which $\vec{j}$ is greater than $\overleftarrow{j}$. At still higher overpotentials $\overleftarrow{j}$ will become negligibly small, and so $j=\vec{j}$ in the region where Tafel's equation is obeyed. It is now supposed, with theoretical support, that Tafel's equation is obeyed throughout the whole range by $\vec{j}$, and likewise by $\overleftarrow{j}$, and that the rounding off of the *net* current density curve at low overpotentials is entirely due to the effect of two opposing reactions. The extrapolation of the Tafel slope therefore represents the values of $\log \vec{j}$ at low overpotentials, and the constant $j_0$ at $\eta=0$ is the value of $\vec{j}$ when the process is in reversible equilibrium. It must also be the value of $\overleftarrow{j}_{rev}$, and $j_0=\vec{j}_{rev}=\overleftarrow{j}_{rev}$ is called the exchange current density. The value obtained will apply to a particular process at a particular concentration.

Figure E.13 shows the data of figure E.12 as an $\eta$–$j$ plot, and also the calculated values of $\overleftarrow{j}$ and $\vec{j}$, assuming that $\alpha=0.5$ for both

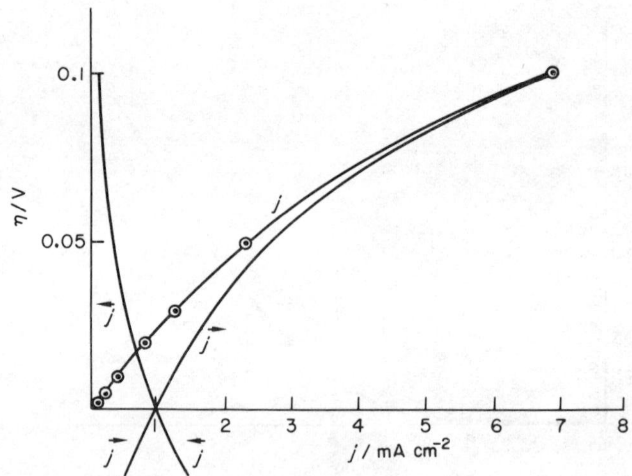


Figure E.13 Tafel diagram for electrode processes. $j$ is the net current density, $\vec{j}$ and $\overleftarrow{j}$ the current densities of the cathodic and anodic currents, respectively

reactions. These two curves intersect at $\eta = 0$, $\vec{j} = j_0$, and the observed current density at any other value of $\eta$ is given by $\vec{j} - \overleftarrow{j}$.

Figure E.14 illustrates another common way of presenting $\eta$–$j$ curves. Potentials are here plotted as abscissae, and the current

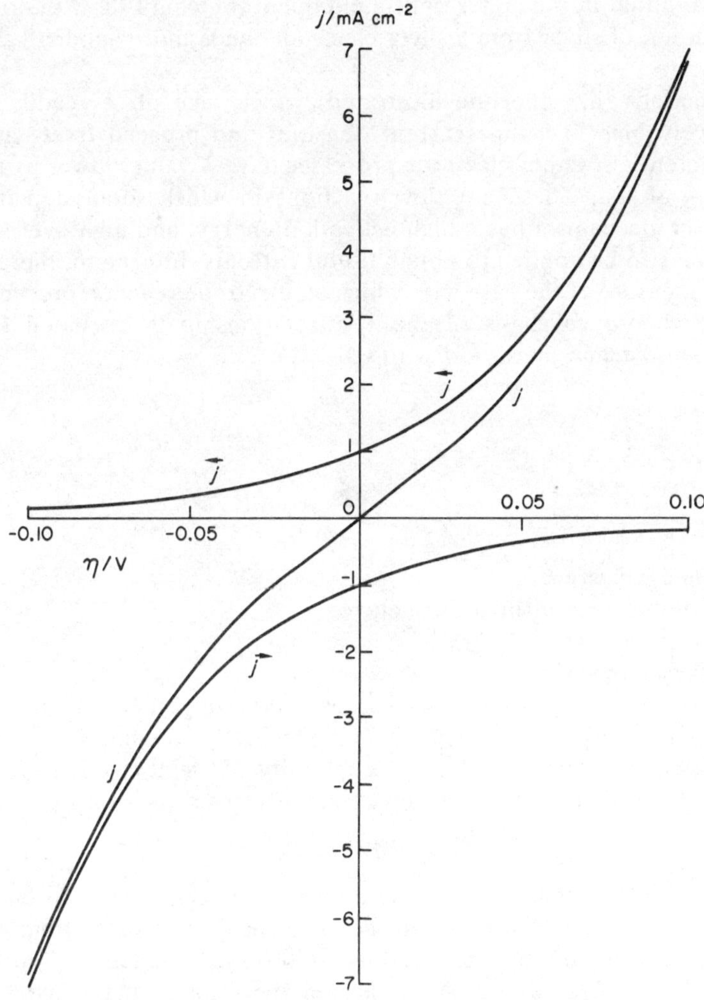


Figure E.14 Current–potential curves for the data of figure E.12

densities as ordinates, and the convention is adopted of showing electron flow from the cathode as a negative current. For instance, if the electrode process under consideration were $Ag^+ + e \rightleftharpoons Ag$, the line in the lower left-hand quadrant would refer to the discharge of silver ions, and that in the upper right-hand quadrant would show the rate of dissolution of silver from a silver electrode made more anodic than the $E^\ominus$ value.

Reactions (like the one illustrated) which take place readily have relatively high exchange current densities and proceed freely at low overpotentials. Other electrode processes have $j_0$ values lower by many powers of ten. These are slow reactions in which some step in the reaction mechanism has a high activation energy, and high overpotentials have to be applied to obtain useful currents. In general, therefore, $j_0$ is a measure of the ease with which an electrode reaction proceeds.

The theoretical basis of the $\eta - j$ relationship is discussed under *electrode reaction mechanisms* (q.v.).

# F

**Falkenhagen effect**
*See* Conductance at high frequencies.

**Faraday constant**
*Faraday's laws* (q.v.) of electrolysis, taken together, mean that a certain quantity of electricity will theoretically discharge one gram-equivalent of any substance. This quantity of electricity is called the Faraday constant or the Faraday, *F*, and is given by

$$F = QE/w = ItA/wz$$

where $Q$ is the quantity of electricity that produces $w$ g of substance of equivalent weight $E$ at an electrode. The charge $Q$ is in coulombs and may be measured by the product $It$. For an element, $E$ may be replaced by $A/z$, where $A$ is the relative atomic mass and $z$ the valency of the element deposited.

The most recent determination of $F$ was made in 1960 by Craig, Hoffman, Law and Hamer, who used a coulometer of the type shown in figure F.1. Three Pyrex beakers connected by syphon tubes contained an electrolyte of 20% perchloric acid and 0.5% silver perchlorate dissolved in water. During electrolysis silver was deposited on the platinum cathode and dissolved from the pure silver anode, and the loss in weight of the anode, after a small correction for insoluble sediment lost by the electrode, was used in the calculation. The relative

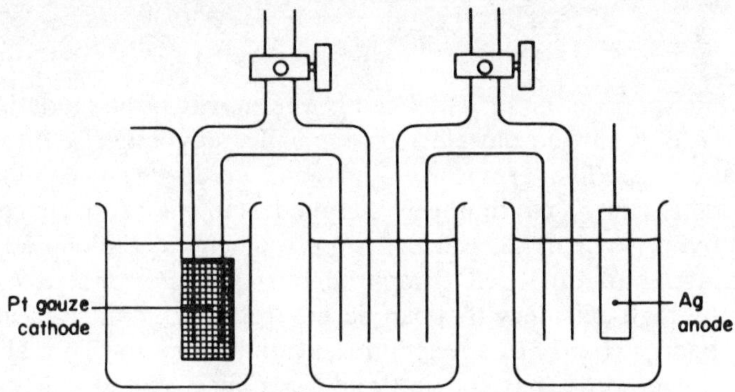


Figure F.1 Silver coulometer of Craig, Hoffman, Law and Hamer

atomic mass of the silver used was also determined, and the value obtained for the Faraday constant was $96\,490 \pm 3\ \mathrm{C\ mol^{-1}}$. A closely agreeing value has been obtained from the *iodine coulometer* (q.v.).

One gram-atom of silver consists of $N_A$ atoms, where $N_A = 6.023 \times 10^{23}\ \mathrm{mol^{-1}}$, the Avogadro constant. The charge on an electron has been accurately determined by physical methods as $1.602 \times 10^{-19}$ C. If the unit ionic charge carried by any univalent ion is the electronic charge, the charge per gram-atom is $6.023 \times 10^{23} \times 1.602 \times 10^{-19} = 9.649 \times 10^4\ \mathrm{C\ mol^{-1}}$, in agreement with the electrolytic value.

**Faraday's laws**

Faraday's laws of electrolytic conduction are:

(1) In any electrolytic process the amount of chemical reaction is proportional to the quantity of electricity passed through the electrolytic conductor.

(2) The masses of different substances deposited or dissolved by the same quantity of electricity are in the proportions of their chemical equivalents.

In ideal cases these laws are exact, although this fact may be obscured in certain cells by the occurrence of side reactions.

**Frequency dependence of conductance**
*See* Conductance at high frequencies.

**Fuel cell**
The popular conception of a fuel cell is one in which the oxidation of a common fuel is carried out electrochemically, so giving rise directly to electrical energy. The term is used, however, to cover any cell in which the reactants can be continuously supplied, and the products continually removed, so that the cell can supply electricity indefinitely. The main attraction of a fuel cell, compared with other power sources, is, of course, the high efficiency that can be theoretically attained in utilising the free-energy change of a reaction as available energy. To realise this in practice requires that the internal resistance of the cell and the *overpotential* (q.v.) at each of the electrode surfaces be low.

So far the most highly developed cell is the hydrogen–oxygen gas cell. The theoretical voltage that can be obtained from the combustion of hydrogen in oxygen is 1.23 V, corresponding to the reaction

$$\begin{array}{r} 2H_2 \rightarrow 4H^+ + 4e \\ O_2 + 2H_2O + 4e \rightarrow 4OH^- \\ 4H^+ + 4OH^- \rightarrow 4H_2O \\ \hline 2H_2 + O_2 \rightarrow 2H_2O \end{array}$$

With efficient catalysts this reaction can give rise at ordinary temperatures to current densities of 100–150 mA $cm^{-2}$ at about 0.75 V. Medium-temperature cells operating at 370–470 K, such as the Bacon cell, can give up to 500–1000 mA $cm^{-2}$ at the same voltage and with less expensive catalysts.

The main problem in design is to bring about the electrode reactions sufficiently rapidly, and at low overpotentials. A part of this problem is solved by using electrode surfaces that catalyse the reaction efficiently. The oxygen reaction, particularly, has a low exchange current density and is normally very slow, and much research into catalytic efficiency has been, and is still being, carried out. The other side to the problem is to bring the gaseous reactants to the electrode–electrolyte interface sufficiently rapidly to maintain reasonably high currents. This is done by using electrodes consisting of thin sheets of porous material—metal, carbon, or metal-coated plastic—and arranging the gas pressure and the pore size so that the gas–liquid interface is within the pores of the material. The liquid meniscus within each pore thus provides an area of electrode wetted by a very thin film of liquid through which the reactant can reach the active surface very readily.

The electrolyte is normally a concentrated solution of a strong acid or alkali; this will have a high conductance and will help to minimise *concentration overpotential* (q.v.). Potassium hydroxide is the first choice as it is less corrosive than concentrated acids, but its usefulness is limited by its reactivity towards carbon dioxide; if used in a hydrogen–air cell, the air must be purified, and for any cell designed to use a hydrocarbon fuel it would be unsuitable as $CO_2$ would be one of the products of the cell reaction. The alternatives are concentrated sulphuric acid, which is only suitable at moderately low temperatures, and phosphoric acid.

The thickness of the electrolyte layer may be as little as 1 mm. In the fuel cells used in the Gemini spacecraft the electrolyte was replaced by an ion-exchange membrane. This provided a conducting bridge between the electrodes, and contributed to the mechanical strength; it also supplied a source of drinking water, since the product of the cell reaction (one pint of water per kilowatt-hour) could be collected in a pure state.

As already indicated, a flat, layer construction is normally used in fuel cell batteries, and an individual cell, usually not more than 1 cm thick, will be somewhat as shown in figure F.2. Low-temperature cells need the most effective catalyst, which is platinum, although silver can be used at the oxygen electrode. To prevent the cost being prohibitive, efforts are made to restrict the platinum coating to the actual region of contact between gas and liquid. In medium-temperature cells, such as the Bacon cell, which operate at about 470 K, nickel is used as catalyst

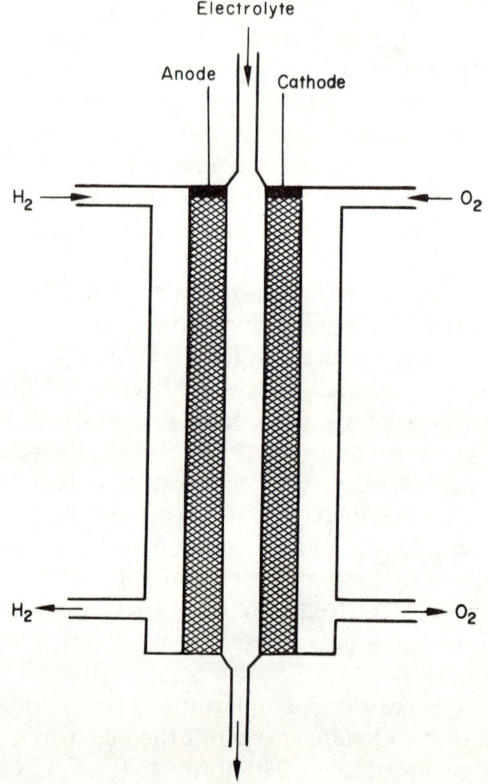


Figure F.2 Hydrogen–oxygen fuel cell (schematic)

at the hydrogen electrode and nickel oxide (on a base of porous nickel) at the oxygen electrode.

The ancillary equipment for such a cell includes pumps, to circulate the gaseous reactants, a condensing unit to remove the product water from the gas streams, and a heating unit to provide the starting temperature (enough heat is generated in the working cell to maintain this). Another practical factor is the source of the hydrogen. Electrolytic hydrogen will not come under consideration, except possibly in electricity storage systems, or where cost is unimportant. It could be produced *in situ* by the action of steam on a cheap liquid fuel, by reactions such as

$$C_3H_8 + 6H_2O \rightarrow 10H_2 + 3CO_2$$
$$CH_3OH + H_2O \rightarrow 3H_2 + CO_2$$

or from liquid ammonia by a catalysed thermal decomposition:

$$2NH_3 \rightarrow N_2 + 3H_2$$

The latter has the advantage that nitrogen is innocuous, whereas $CO_2$ might have to be removed.

There would obviously be great advantages in using an organic fuel directly in the cell, and avoiding the subsidiary reactions just mentioned. Methanol or propane will react anodically, thus

$$C_3H_8 + 6H_2O \rightarrow 20H^+ + 3CO_2 + 20e$$
$$CH_3OH + H_2O \rightarrow 6H^+ + CO_2 + 6e$$

The difficulty is that these reactions are slow, and their useful application requires expensive catalysts and elevated temperatures. Also, an acid electrolyte must be used since $CO_2$ is a product, and the formation of intermediate products such as formaldehyde must be prevented to avoid poisoning the catalyst. At present the disadvantages in the direct use of such fuels appear to be greater than those of employing them indirectly to provide the fuel for a Bacon cell.

High-temperature fuel cells using a fused salt as electrolyte, or a conducting oxide, are also being studied. For instance, a hydrocarbon fuel may be used with a fused mixture of alkali metal carbonates in the temperature range 770–970 K. The fuel gives with steam a mixture of carbon monoxide and hydrogen at the operating temperature, and the anode reactions are

$$2CO + 2CO_3^{2-} \rightarrow 4CO_2 + 4e$$
$$2H_2 + 2CO_3^{2-} \rightarrow 2H_2O + 2CO_2 + 4e$$

The cathode reaction is

$$O_2 + 2CO_2 + 4e \rightarrow 2CO_3^{2-}$$

so that the over-all reactions are

$$2CO + O_2 \rightarrow 2CO_2$$
$$2H_2 + O_2 \rightarrow 2H_2O$$

*See also* Fr; and Liebhafsky, H. A. and Cairns, E. J. (1968), *Fuel Cells*, Wiley.

**Fuoss conductance equation**
*See* Conductance equations.

**Fused salts**
As electrolytes, fused salts were first studied by Faraday, and are now attracting much interest as high-temperature solvents.

Conductance results are more difficult to compare than those in water because of wide differences in temperature and viscosity; this difficulty has been partly overcome by comparing the conductance of salts at temperatures 10 K above their respective melting points, regarding these as 'corresponding temperatures'. Typical salts, such as the alkali metal chlorides, show molar conductivities (conductivity × molar volume) of the same order of magnitude at their melting points as for their aqueous solutions. There is no doubt, therefore, that their melts consist essentially of free ions, and constitute a new type of solvent in which interionic forces are very high. The results for the alkali metal chlorides differ from those in aqueous solutions, however, in that the conductivity is highest for the lithium salt, and falls off with increasing crystal radius. This is in keeping with the supposition that in the absence of solvent it is the bare ion that moves in the applied field. The alkaline earth chlorides are also good conductors, but the order is now reversed: $MgCl_2$ has less than half the conductivity of $BaCl_2$, and $BeCl_2$ is a weak electrolyte. This effect can probably be attributed to ion pairing, and on passing further to the right in the periodic table, this influence of increasing valency is reinforced by an increasing tendency to covalent bonding, so that the well-conducting chlorides are confined to a triangle in the lower left-hand corner of the table, while the chlorides of Al, Ti, Ta and W are virtually non-conductors.

*Transport numbers*

In a molten salt these cannot be determined in the same way as in a normal solvent. Consider, for instance, a cell with silver electrodes and fused silver nitrate as electrolyte. When 1 $F$ of electricity is passed, 1 g-equiv. of silver is deposited, and if $t_c$ equivalents are provided by the migration of silver ions, the remaining $(1-t_c)$ equivalents are provided by the movement of $AgNO_3$ towards the cathode.

Corresponding changes would occur at the anode, and no concentration gradients result; the frame of reference provided by a stationary solvent is absent. Various methods have been tried for measuring the mass movement of the liquid, or its effects, but until these are better established it may be better to derive mobility values from diffusion experiments. The individual ions can be studied in, say, NaCl by separate self-diffusion measurements using, respectively, $^{22}Na$ and $^{36}Cl$ as radiotracers. From these results for motion under a known concentration gradient the mobility under unit potential gradient can be calculated. A correction for ion-pairing is required, since ion-pairs will contribute to the diffusion, but not to the electrical transport. This correction follows from the amount by which the observed conductivity of the salt falls short of that calculated from the diffusion results.

Mixtures of molten salts have the advantage of lower melting points, and comparative results can be obtained by using one salt as the main constituent—the solvent—to which smaller amounts of a second substance are added. For mixtures of the salts of the alkali metals, the conductances are high and approximately additive. When an alkali metal chloride is mixed with a poorly conducting chloride, such as those of Zn, Cd or Al, there is evidence that the latter takes up chloride ions to form a complex anion. This is in keeping with the tendencies observed in aqueous solutions, and complex-ion formation, like ion-pairing, can be expected to be common in ionic melts.

Metals may dissolve in fused salts, and may interact to give metal ions in a subnormal valency state, complications in the electrowinning of metals. The conductance of the metal solution is high and largely electronic, and Faraday's laws will not apply to the melt.

The use of fused salts in electrometallurgy (Al, Na, etc.) is well known. Mixtures of fused salts may also be used in high-temperature cells for power production, where they enable high current densities and low internal resistance to be attained.

*See also Specialist Periodical Reports, Electrochemistry,* **3** (1972) (The Chemical Society, London).

# G

**Galvanic cell**

*See* Reversible galvanic cell

**Galvanostat**

*See* Amperostat.

**Gas electrode**

The gas electrode is an *electrode* (q.v.) in which one part of the electrode couple is a gas, e.g. hydrogen or chlorine. A chemically inert conductor, usually platinised platinum, is used to adsorb the gas, to transport electrons and to catalyse the electrode reaction.

The *electrode potential* (q.v.) depends on the pressure of the gas and the activity of the ions in solution. Thus, for the chlorine electrode,

$$E(\mathrm{Cl_2, Cl^-}) = E^{\ominus}(\mathrm{Cl_2, Cl^-}) + \frac{RT}{F} \ln \frac{p^{1/2}(\mathrm{Cl_2})}{a(\mathrm{Cl^-})}$$

This, in combination with a hydrogen electrode, gives the cell

$$\ominus \quad \mathrm{Pt, H_2} \,|\mathrm{H^+ \;\; Cl^-}|\, \mathrm{Cl_2, Pt} \quad \oplus$$

for which the reaction, for the passage of 1 Faraday of electricity, is

$$\tfrac{1}{2}\mathrm{H_2} + \tfrac{1}{2}\mathrm{Cl_2} \rightleftharpoons \mathrm{H^+} + \mathrm{Cl^-}$$

in which the HCl is formed at the molality existing in solution:

$$E(\text{cell}) = E(\mathrm{Cl_2, Cl^-}) - E(\mathrm{H^+, H_2})$$

$$= E^{\ominus}(\mathrm{Cl_2\, Cl^-}) - \frac{2\,RT}{\mathrm{F}} \ln a_{\pm} + \frac{RT}{2\mathrm{F}} \ln \frac{p(\mathrm{Cl_2})}{p(\mathrm{H_2})}$$

*See also* De, I & J.

**Glass electrode**

The e.m.f. of the cell

$$\underbrace{\text{Ag, AgCl} \left| \begin{matrix}\text{HCl} \\ (0.1\ \text{mol dm}^{-3})\end{matrix} \right| \text{Glass}}_{\text{Glass electrode}} \left| \begin{matrix}\text{Test} \\ \text{solution}\end{matrix} \right\vdots \begin{matrix}\text{KCl} \\ \text{sat.}\end{matrix} \left| \ \text{Hg}_2\text{Cl}_2\text{, Hg} \right.$$

varies with the hydrogen ion concentration of the test solution. The glass electrode may therefore be used for the determination of *pH* (q.v.); its electrode potential may be represented by the equation

$$E_G = E'_G + \frac{RT}{F} \ln a(H_3O^+)$$

Modern glass electrodes consist of a bulb of special glass (72% $SiO_2$, 8% CaO, 20% $Na_2O$) blown on the end of ordinary glass tubing (figure G.1). The bulb, when supplied, contains hydrochloric acid (0.1 mol $dm^{-3}$) with a silver–silver chloride electrode dipping in it. The only preparation required is that the bulb must be soaked in hydrochloric acid (0.1 mol $dm^{-3}$) for 24 hours before use. These electrodes should not be used in solutions of pH > 11; electrodes of special glass are available for use from pH 9 to 14. Glass electrodes must never be wiped or allowed to dry; they should be stored in distilled water.

The glass electrode, with a resistance of $10^7$–$10^8$ $\Omega$, cannot be used with simple potentiometer circuits; a *pH meter* (q.v.) with a valve or transistor circuit is necessary. In use, the glass electrode and its reference electrode must be calibrated using solutions of known pH values (Table A.V, p. 245), to eliminate asymmetry potentials in the glass.

It has an accuracy of ±0.05 pH, or better if extreme care is taken, over the range pH 2–10; it reaches equilibrium immediately in any solution and is unaffected by the presence of any gas, oxidising and reducing agents, and poisons in the generally accepted sense; it has no appreciable salt or protein error; it can be obtained in many forms for pH measurements in solutions, emulsions or pastes, and for measurements on a micro scale (0.02 $cm^3$); if carefully handled and stored in distilled water the electrode does not deteriorate with age.

Owing to its high electrical resistance, the glass electrode can only be used with a valve or transistor potentiometer; the electrode suffers

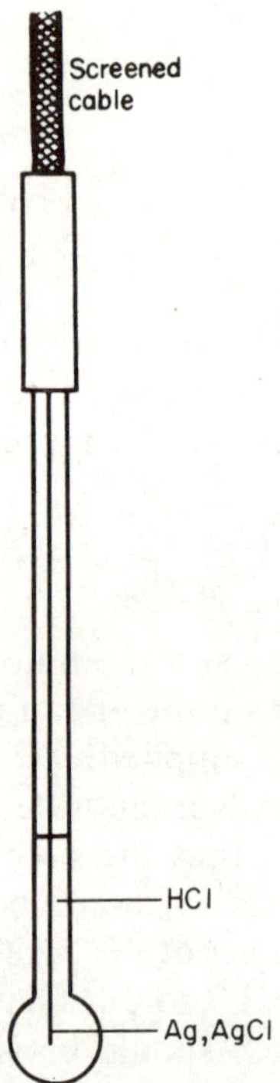


Figure G.1 Glass electrode

from asymmetry potentials, thus necessitating regular standardisation; it is very sensitive to previous treatment and should be well washed after use. After use in the alkaline region it is advisable to wash it with dilute acid and water; owing to its extreme sensitivity to electrical interference, the wire connection to the pH meter must be screened and earthed.

*See also* B, G, I & J.

**Gouy layer**

*See* Electrical double layer

# H

**Half-cell**
*See* Electrode.

**Half-wave potential**
*See* Polarography.

**Helmholtz layer**
*See* Electrical double layer

**Hittorf method**
The analytical method of measuring transport numbers was first used by Hittorf in 1857. Consider the cell shown diagrammatically in figure H.1, in which a silver nitrate solution is electrolysed between silver electrodes. By *Faraday's laws* (q.v.), when $x$ Faradays of electricity are passed, $x$ gram-equivalents of silver dissolve from the anode as ions, and $x$ gram-equivalents of silver ions are discharged and deposited on the cathode; 96 490 $x$ coulombs of negative electricity, released at the anode, will have passed through the external circuit to the cathode, where they will have neutralised silver ions. The circuit is completed by electrolytic conduction through the solution; silver ions migrate towards the cathode and nitrate ions carry negative charges

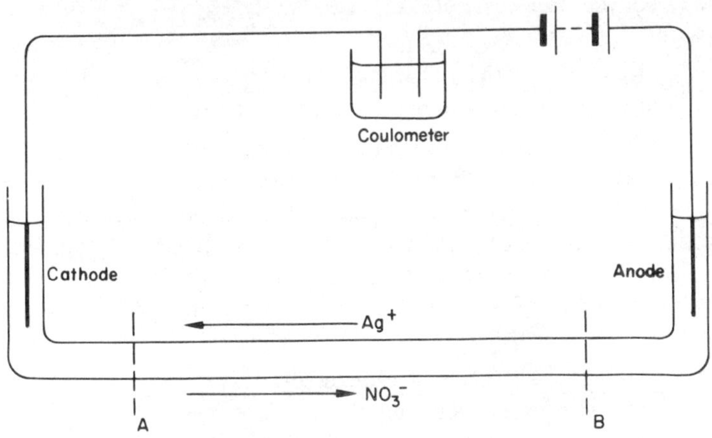


Figure H.1 Transport number measurement (diagrammatic)

towards the anode, and the sum of these effects will be the transfer of 96 490 $x$ coulombs through the solution.

With the passage of current, the solution round the anode gains in silver nitrate content and the solution round the cathode loses silver nitrate. These concentration changes move out progressively from the electrode surfaces, but (provided that the current has not been passed for too long) the greater part of the intermediate solution remains unchanged in composition, since ions migrating out of a given volume are replaced by an equal number moving in. A cross-section A may therefore be chosen at some point where the concentration is unchanged at the end of the experiment, the whole loss of $AgNO_3$ round the cathode being confined to the 'cathode portion' of solution to the left of A. The charge of $x$ Faradays is carried across plane A partly by $NO_3^-$ moving from left to right, and partly by $Ag^+$ moving in the opposite direction. Their numbers are equal and they carry equivalent charges, but their velocities may differ (reflecting differences in size and in the resistance they encounter from the solvent). If these are written $u$ for the cation and $v$ for the anion, the fraction of the current carried across A by $Ag^+$ will be $u/(u+v)$; i.e. $ux/(u+v)$ g-equiv. of $Ag^+$ will have entered the cathode portion during the experiment and $vx/(u+v)$ g-equiv. of $NO_3^-$ will have been lost by it. Simultaneously $x$ g-equiv. of $Ag^+$ will have been discharged and deposited at the cathode, so the net loss of $Ag^+$ is also $\{x-ux/(u+v)\}=vx/(u+v)$. Therefore,

$$\frac{\text{loss of AgNO}_3\text{ at cathode in g-equiv.}}{\text{quantity of electricity passed in Faradays}}=\frac{v}{u+v}=t_a$$

where $t_a$ is the *transport number* (q.v.) of the anion.

An analysis of the anode portion could be used equally well. Here $x$ g-equiv. of $Ag^+$ will have entered the solution by dissolution at the electrode, but $ux/(u+v)$ g-equiv. will have passed across plane B from right to left. The anode portion will thus have gained $\{x-ux/(u+v)\}=vx/(u+v)$ g-equiv. of $Ag^+$ and an equal amount of $NO_3^-$ will have entered across plane B. Therefore,

$$\frac{\text{gain of AgNO}_3\text{ at anode in g-equiv.}}{\text{quantity of electricity passed in Faradays}}=\frac{v}{u+v}=t_a$$

The cation transport number follows from $t_c=1-t_a$.

The same arguments can be applied to any electrolysis, but the resulting formula will depend on the electrode reactions. Suppose that potassium hydroxide is being electrolysed with platinum electrodes. The electrode reactions are

$$\text{at cathode}\quad K^+ + H_2O + e \rightarrow K^+ + \tfrac{1}{2}H_2 + OH^-$$

$$\text{at anode}\quad 2OH^- \rightarrow H_2O + \tfrac{1}{2}O_2 + 2e$$

The changes occurring in the electrode regions, per Faraday, are

| Cathode portion | | Anode portion |
|---|---|---|
| 1 g-equiv. $OH^-$ formed and | | 1 g-equiv. $OH^-$ lost and |
| $v/(u+v)$ g-equiv. $OH^-$ lost | $\xrightarrow{OH^-}$ | $v/(u+v)$ g-equiv. $OH^-$ gained |
| $u/(u+v)$ g-equiv. $K^+$ enter | $\xleftarrow{K^+}$ | $u/(u+v)$ *g-equiv.* $K^+$ leave |

Hence,

$$\frac{\text{gain of KOH at cathode}}{\text{Faradays passed}} = \frac{u}{u+v} = \frac{\text{loss at anode}}{\text{Faradays passed}} = t_c$$

In this derivation of transport number formulae the exact distribution of concentration changes near the electrodes need not be known; they should be understood, however. If the ions travelled at equal speeds, exactly half of the silver deposited in the first example would have migrated into the cathode chamber. The other half comes from the silver nitrate in the solution close to the cathode, which therefore falls in concentration (nitrate ions also being lost by migration towards the anode). Salt now diffuses down the concentration gradient from the unchanged solution towards the electrode surface, and this diffusion provides the other half of the silver deposited (and maintains the migration of nitrate ions away from the cathode). For a steady current, the concentration gradient will increase until it is just sufficient to cause diffusion at the necessary rate.

Note that the derivation is based on the passage of ions in and out of fixed 'electrode portions' determined by the planes of reference A and B, and the gain or loss of electrolyte is calculated for a constant quantity of solvent. Appreciable volume changes at the electrodes would affect this picture, and are allowed for in very accurate work. A net movement of the solvent under the influence of the current would also interfere, and this could occur if one of the ions carried more solvent with it than the other. Attempts have been made to determine 'true' transport numbers which would correct for this effect, by adding a non-electrolyte as a reference substance, and observing any small change of concentration of this near the electrodes. If the reference substance is assumed not to move, a net movement of water would be detected in this way. These attempts have been found to be invalid, however, as the results vary with the reference substance used.

*Experimental arrangement*
The apparatus is designed with the following considerations in mind.

(a) The apparatus must be firmly supported and vibration-free, to minimise mixing.

(b) If gases are produced, the electrodes must be at the top of the column of electrolyte, and this must be withdrawn in such a way that the liquid of changed composition is all washed out by liquid of unchanged composition.

(c) If the solution at the electrode becomes more concentrated and denser, the electrode should be at the bottom of the column of liquid; if less dense, at the top.

(d) The current must not be high enough to cause appreciable heating, or convection currents will cause mixing; it must not be passed for so long that diffusion has time to affect the intermediate solution.

(e) Constrictions and taps in the path of the current must be avoided, or local heating will cause mixing.

A common form of apparatus is shown in figure H.2. The cell is filled with electrolyte solution, and a current of 10–20 mA is passed for 1–2 h. Intermediate portions are withdrawn from B and C, and should be unchanged in composition. One or both electrode portions are then withdrawn through A and D, weighed and analysed.

Suppose that a silver nitrate solution containing 0.0847 g $AgNO_3$ in 10.058 g solution had been electrolysed between Ag electrodes, and after electrolysis the anode portion of 27.04 g contained 0.2818 g $AgNO_3$, while 0.0194 g copper was deposited on the coulometer cathode. The starting point of the calculation is the anode portion which

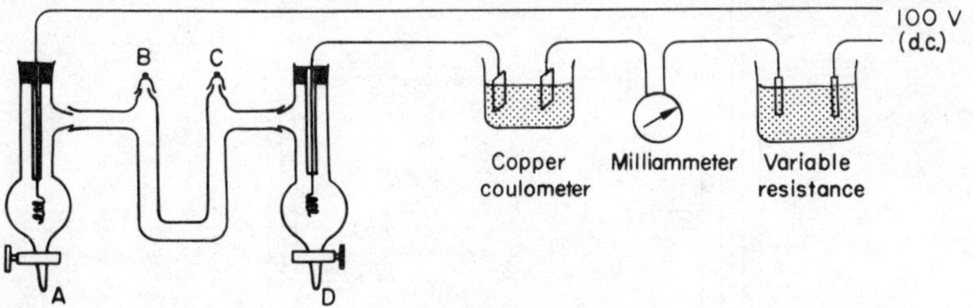


Figure H.2 Hittorf transport number apparatus

contained 0.2818 g $AgNO_3$ and 26.76 g $H_2O$. Before electrolysis this water contained $26.76 \times 0.0847/9.973 = 0.2273$ g $AgNO_3$. The gain of $AgNO_3$ is therefore $0.0545/170 = 0.000321$ g-equiv. The quantity of electricity passed is $0.0194/31.79 = 0.000\,610\,F$, and $t_a = 321/610 = 0.526$, and $t_c = 1 - t_a = 0.474$.

*See also* De, J & P.

**Hydration of ions**
*See* Molar ionic conductivity.

**Hydrogen electrode**
When *platinum* (q.v.) coated with platinum black is saturated with hydrogen gas, and immersed in a solution containing $H^+$, it behaves like a metallic electrode:

$$\tfrac{1}{2}H_2\,(g) \rightleftharpoons H^+ + e$$

The *electrode potential* (q.v.) is given by

$$E(H^+, H_2) = E^{\ominus}(H^+, H_2) + \frac{RT}{F} \ln \frac{a(H^+) \times (101\,325)^{1/2}}{(p(H_2)/N\,m^{-2})^{1/2}}$$

It is the recognised convention to take the standard hydrogen electrode, in which $p(H_2) = 1$ atm ($101\,325\,N\,m^{-2}$) and $a(H^+) = 1$, as the arbitrary zero of electrode potential. Thus $E^{\ominus}(H^+, H_2) = 0$, and, hence,

$$E(H^+, H_2) = \frac{RT}{F} \ln \frac{a(H^+) \times (101\,325)^{1/2}}{(p(H_2)/N\,m^{-2})^{1/2}}$$

The simple form of the hydrogen electrode (figure H.3) consists of a

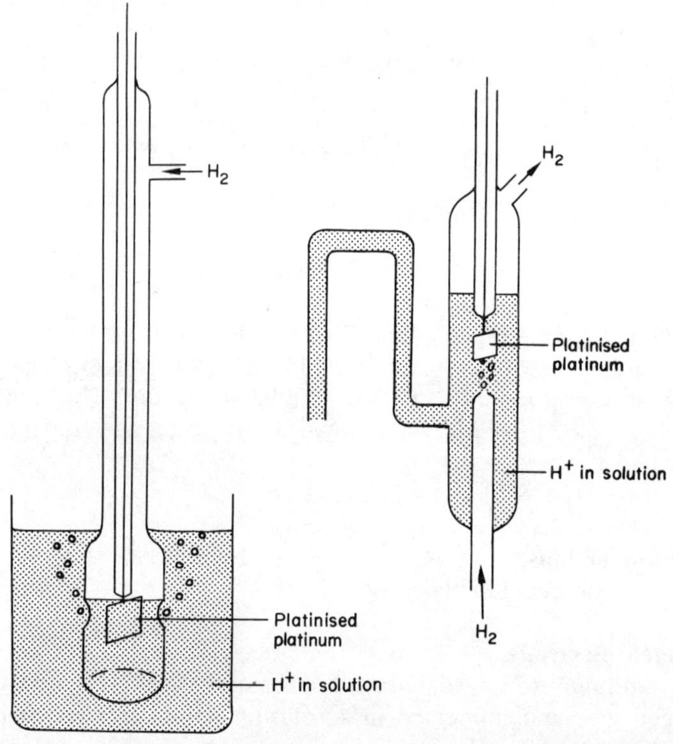


Figure H.3 Typical forms of hydrogen electrode

piece of platinised platinum (area 2 cm$^2$) spot welded to a piece of platinum wire and fused into soft glass tubing which carries the electrical connection. Hydrogen is passed through the wider tube and is bubbled into the test solution.

Cylinders of hydrogen provide the most convenient source of the gas, which must be passed through wash bottles containing a solution of alkaline pyrogallol to remove oxygen, dilute sulphuric acid, water and a sample of the solution under test in the cell (this latter prevents alteration of the concentration of the solution in the cell).

In operation a steady stream of hydrogen is bubbled through the solution until the cell (comprising the hydrogen electrode and a reference electrode) assumes a constant e.m.f. This may take 10–15 min; thereafter the e.m.f. is independent of the rate of bubbling.

The advantages of the hydrogen electrode are: it is capable of a high degree of accuracy, giving reproducible results over the complete pH range 0–14; there is no salt error, i.e. no apparent shift of pH caused by variation of the ionic strength; the electrical resistance of the electrode is small, and it can therefore be used with a normal potentiometer.

The limitations are: it cannot be used in the presence of air, oxygen or oxidising and reducing reagents; the platinum black deteriorates and must be frequently renewed; the platinised surface is readily poisoned by alkaloids, cyanides, arsenic and antimony compounds, and by colloids which are adsorbed on the surface.

*See also* G, I & J.

**Hydrogen evolution reaction**

Since electrode potentials are quoted on the hydrogen scale, it might be expected that only metals with positive standard electrode potentials could be deposited from acid solutions, and that hydrogen would be discharged in preference to other cations, such as lead ($E^{\ominus}$= −0.126 V) or nickel ($E^{\ominus}$= −0.25 V). The failure of this prediction is due to the high *activation overpotential* (q.v.) of hydrogen on many metals, values obtained at a current density of 1 mA cm$^{-2}$ varying from 0.01 V on a platinised platinum cathode to as much as 0.67 V on lead and 1.04 V on mercury.

Hydrogen is discharged from acid solutions by the over-all reaction

$$2H_3O^+ + 2e \rightarrow H_2 + 2H_2O$$

and from alkaline solutions by the reaction

$$2H_2O + 2e \rightarrow H_2 + 2OH^-$$

Acid solutions have had most study, and the steps involved in this case can be visualised as follows.

(a) Hydrogen ions from the bulk of the solution move up into the outer plane of the *electrical double layer* (q.v.).

(b) The hydrogen ion is transferred across the double layer, receiving an electron and becoming dehydrated, to give a hydrogen atom adsorbed on the electrode surface.

(c) Hydrogen molecules are formed. The obvious mechanism is (*i*) the process $2H \rightarrow H_2$ at the metal surface, but a second possibility is (*ii*) the arrival of a second hydrogen ion at a site at which a hydrogen atom is adsorbed: $H + H_3O^+ + e \rightarrow H_2 + H_2O$.

(d) The hydrogen molecules are desorbed.

(e) Bubbles are formed and hydrogen gas is evolved.

Some of these steps can be eliminated as of minor concern only. If step (a) were to become the rate-controlling factor in an experiment, this could easily be recognised as a *concentration overpotential* (q.v.). Step (d) will be unimportant, because once molecular hydrogen has been formed, this is known to be easily desorbed from metal electrodes. Step (e) has had some attention, and small fluctuations in potential have been observed at a point cathode as the bubbles disengage and fresh ones form; the effect is relatively quite unimportant, however.

The centre of interest is, therefore, whether the reaction goes by mechanism (c, *i*) or (c, *ii*), and whether this or (b) is the rate-controlling step. The conclusions that have been reached are that on mercury, lead and cadmium—high-overpotential metals—the discharge of protons is the rate-determining step, and is followed by (c, *ii*). On other metals, proton discharge takes place more readily, and is not the slow step. For platinum and rhodium this is step (c, *i*); for metals with intermediate values for the overpotential, such as nickel and gold, proton discharge is followed by (c, *ii*), and the latter step is normally rate-controlling.

The chief methods by which these conclusions have been reached are summarised under *electrode reaction mechanisms* (q.v.). First, the $\eta$–$\log j$ curve gives the value of the 'Tafel slope' $b$, and, hence, the transfer coefficient $\alpha$, since $b = 2.303\, RT/\alpha F = 0.059/\alpha$ (see *activation overpotential*). If the rate is determined by the proton discharge step, $\alpha$ will be simply the symmetry factor $\beta$, and if this is approximately 0.5, $b$ will have the value 0.118. This is what is found for the mercury group.

If the process (c, *i*) is rate-controlling, protons will be readily discharged, and a monolayer of adsorbed atoms will build up on the surface to a steady state value where the number of atoms arriving at the surface is balanced by the sum of the number combining to form molecules and the number re-ionising. This situation will arise at a metal where the *exchange current density* (q.v.) is high, and the processes $H_3O^+ + e \rightleftharpoons H + H_2O$ can be regarded as virtually in equilibrium. If $\theta$, the fraction of surface covered, is taken as a measure of the activity of adsorbed hydrogen atoms, this equilibrium condition can be expressed by the equation

$$\underset{\text{discharge current}}{k_1[H_3O^+](1-\theta)\exp(\beta F\eta/RT)} = \underset{\text{ionisation current}}{k_2\theta\exp\{(1-\beta)F\eta/RT\}}$$

so that

$$\theta/(1-\theta) = (k_1[H_3O^+]/k_2)\exp(-F\eta/RT)$$

If the surface coverage is small, this can be approximated to

$$\theta = k_1 k_2^{-1}[H_3O^+]\exp(-F\eta/RT)$$

The velocity of the rate-determining step is $v = k_3\theta^2$, and so the current will be given by

$$j = Fk_3(k_1 k_2^{-1}[H_3O])^2 \exp(-2F\eta/RT)$$

and the Tafel slope will be four times smaller than in the previous case.

Process (c, *ii*) implies that hydrogen atoms reach the surface by proton discharge, and leave it either by ionising or by combining with a proton. The rates of these three processes can be written:

$$\text{rate of discharge} = k_1[H_3O^+](1-\theta)\exp(-\beta F\eta/RT) \quad \text{(H.1)}$$
$$\text{rate of ionisation} = k_2\theta\exp\{(1-\beta)F\eta/RT\} \quad \text{(H.2)}$$
$$\text{rate of combination} = k_3\theta[H_3O^+]\exp(-\beta F\eta/RT) \quad \text{(H.3)}$$

This mechanism might apply to a metal which is a poor catalyst for the process $2H_{ads} \rightarrow H_2$, and which has a low exchange current density. A high coverage of $H_{ads}$ would first be established, and then $H_2$ would be formed by reaction (H.3). When the approximation $\theta \approx 1$ is applied to reaction (H.3), the result is to give an equation indistinguishable from that of the first mechanism considered above, with $\alpha = \frac{1}{2}$ and $b = 0.118$. However, for a metal with a higher exchange current density, reactions

(H.1) and (H.2) might be regarded as virtually in equilibrium:

$$k_1[H_3O^+](1-\theta)\exp(-\beta F\eta/RT) = k_2\theta\exp\{(1-\beta)F\eta/RT\}$$

For $(1-\theta)\approx 1$, this would give

$$\theta = k_1 k_2^{-1}[H_3O^+]\exp(-F\eta/RT)$$

and this value in reaction (H.3) would yield

$$j = Fk_1k_2^{-1}k_3[H_3O^+]^2\exp\{-(1+\beta)F\eta/RT\}$$

With $\beta = \frac{1}{2}$, this mechanism would give the distinctive values $\alpha = \frac{3}{2}$, $b = 0.039$, which have been found for some metals.

The evidence from Tafel slopes can be supplemented in a number of ways. The assumptions made about the coverage of the electrode have been tested by abruptly changing the potential of the working electrode to a value at which the adsorbed hydrogen is re-ionised. An integral of the current–time curve over the brief interval in which the adsorbed film is removed gives a value for the hydrogen coverage. Isotope effects have also given valuable information. Theoretical models for the transition states corresponding to different rate-determining steps give a spread of values for the hydrogen–tritium separation factors with which experimental values can be compared.

*See also* B & R.

**Hydrogen overpotential**

*See* Activation overpotential.

# I

**Ilkovic equation**

*See* Polarography.

**Indicator**

Indicators which respond with a colour change to change of pH or of oxidation potential are used to locate the end-point in a titration.

*Acid–base indicators*

The pH of the exact equivalence point in an acid–base titration depends on the relative strengths of the acid and the base. For the titration of a strong acid with a strong base it is at pH 7.0 (neutrality), while it is on the acid (alkaline) side of neutrality if the base (acid) is weak, owing to the hydrolysis of the salt. The indicator for a given titration must exhibit its colour change as close as possible to the pH of the equivalence point.

Acid–base indicators are generally weak acids which possess different colours according to the pH of the solution. Classically,

$$HIn \rightleftharpoons H^+ + In^-$$

for which the apparent ionisation constant (neglecting activity coefficients) is

$$K_{In} = [H^+][In^-]/[HIn]$$

or

$$pH = pK_{In} + \log [In^-]/[HIn]$$

Thus, when the indicator is added to a solution of given pH, the equilibrium adjusts itself until these equations are obeyed. Since the two forms of the indicator, HIn and $In^-$, have different colours, the colour of the indicator and, hence, of the solution is therefore adjusted. From the ratio of the intensities of the two colours the pH of the solution may be determined.

On account of the difficulty of detecting a small intensity of one colour in the presence of another, the useful range of an indicator is limited to

$$pH = pK_{In} \pm 1$$

i.e. from $[In^-]/[HIn] = 10$ to $[HIn]/[In^-] = 10$. This enables the correct indicator to be chosen for the particular titration (Table A.VI, p. 246).

$K_{In}$ may be determined by methods used for the determination of the *dissociation constant*‡ (q.v.) of an acid, particularly the spectrophotometric method.

*Oxidation–reduction indicators*

The indicator used to determine the equivalence point in an oxidation–reduction titration should have an oxidation potential midway between the redox potentials of the two reagents in the titration, and must exhibit a sharp colour change. These indicators, generally organic dyestuffs, have different colours in the oxidised and reduced forms. Like acid–base indicators, their range is also limited by the intensity of colours that can be distinguished, i.e. from $c(\text{red})/c(\text{ox}) = 10$ to $c(\text{ox})/c(\text{red}) = 10$, or

$$E = E^{\ominus}_{\text{In}} \pm 0.06$$

For a redox titration to be quantitative, the standard redox potentials of the two systems must differ by at least 0.3 V. The rapid change of potential at the end-point should thus cover a range of at least 0.12 V. It is possible to achieve a very sharp end-point with a redox indicator provided that its working range is covered by the change of potential at the end-point (Table A.VII, p. 246).

To overcome some of the limitations of visual indicators, e.g. in highly coloured solutions, instrumental methods have been developed to locate the end-point. *Electrometric titrations* (q.v.) are extensively used.

*See also* J & P, V.

**Indicator electrode**

The definition of suitable indicator electrodes forms the basis of both theoretical and practical aspects of e.m.f. measurements, e.g. for obtaining basic thermodynamic data, elucidation of electrode processes and *in situ* analyses. The three that meet the requirements most adequately are the *hydrogen electrode* (q.v.), the *silver–silver chloride electrode* (q.v.) and the *calomel electrode* (q.v.). The only electrode to which the term 'reference electrode' may be rigorously applied is the hydrogen electrode. Of the secondary electrodes, a distinction is drawn between those for which the standard potential can be expressed in terms of strictly thermodynamic quantities and those which are less easily explained on purely thermodynamic principles, e.g. the *glass electrode* (q.v.).

Some applications of different electrodes are tabulated on p. 147.

**Indifferent electrolyte**

An indifferent electrolyte is a constituent of an electrolyte solution that takes no part in the electrode processes under study. Its functions are

| Electrode | Suitability: aqueous solutions | organic solvents | biological systems | Special applications and reservations |
|---|---|---|---|---|
| Hydrogen | 4 | 2 | 1 | Electrode standardisation |
| Silver–silver halide | 4 | 3 | 3 | Secondary reference electrode, thermodynamic studies |
| Calomel | 4 | 2 | 2 | Thermodynamic studies (contaminates solution) |
| Glass | 4 | 2 | 4 | pH measurements, potentiometric titrations |
| Quinhydrone | 4 | 1 | 1 | Potentiometric studies (contaminates solution) |
| Metal–metal oxide | 3 | 1 | 1 | Best used in alkaline range |
| Oxygen | 2 | 1 | 3 | Measurements of oxygen tension |

4, generally applicable.
3, applicable to selected systems.
2, occasionally suitable.
1, inapplicable or insufficient data information available.

to reduce the resistance of the solution and also, by carrying almost all the current, to ensure that the electroactive constituents (those taking part in the electrode reactions) reach the electrode surfaces by diffusion and not by electrolytic transport.

**Internal electrolysis**

In internal electrolysis, which is a variant of *electrogravimetric analysis* (q.v.) or *coulometry* (q.v.) the substance to be estimated is made part of a galvanic cell. For instance, the copper(II) ion content of a solution can be estimated by immersing in it electrodes of Pt and Zn. When these are connected through the external circuit, the Daniell cell reaction, $Cu^{2+} + Zn \rightarrow Zn^{2+} + Cu$, will take place, zinc ions going into solution, and copper being deposited on the Pt electrode. The current can be controlled by a variable resistance, and the solution should be stirred. When all the copper has deposited, hydrogen will be evolved, and the cathode is washed and weighed.

If the products of an internal electrolysis are soluble, the quantity of electricity, instead of a gain in weight, will be measured, and Faraday's law then gives the number of moles reacted.

**Iodine coulometer**
This very accurate coulometer makes use of the reaction $e+\frac{1}{2}I_2 \rightleftharpoons I^-$. The cathode and anode compartments contain Pt electrodes, and are separated by a column of 10% KI solution; a standardised solution of $I_2$ in KI is introduced to cover the cathode. During electrolysis, iodine is formed at the anode and dissolves in the KI solution, while iodine is reduced to iodide at the cathode. After the experiment both the loss and gain can be determined volumetrically.

The reaction is well adapted to measuring very small quantities of electricity. In this case the iodine formed at the anode is determined in a spectrophotometer against standard iodine solutions ($\pm 1$ mg $I_2 \equiv$ 0.76 C).

**Ionic atmosphere**
*See* Conductance equations.

**Ionic melts**
*See* Fused salts.

**Ionic mobility**
*See* Molar ionic conductivity

**Ion-pair**
A completely dissociated electrolyte should obey the Debye–Hückel equation and *Onsager's equation* (q.v.) in sufficiently dilute solutions. Many salts in water conform with this expectation, but some uniunivalent and unidivalent salts, and most of those of higher valence type, show abnormally low molar conductivities and activities. The deviations can be explained by association between cation and anion to form ion-pairs, and the equilibrium between free ions and ion-pairs can be formulated, in the same way as for a weak electrolyte, by writing $K_D = \gamma_C\gamma_A\alpha^2 m/(1-\alpha)$, where the $\gamma$s are the activity coefficients of cation and anion, $\alpha$ is the fraction of electrolyte present as free ions and $K_D$ is the dissociation constant. The reciprocal of this, $K_A$, the association or stability constant, is also used, and it will be seen that $pK_A = -pK_D$. The ion-pair itself may carry a net charge, e.g.

$Pb^{2+} + 2Cl^- \rightleftharpoons PbCl^+ + Cl^-$, and then

$$K_D = \frac{\gamma(Pb^{2+})\gamma(Cl^-)}{\gamma(PbCl^+)} \times \frac{m(Pb^{2+})m(Cl^-)}{m(PbCl^+)}$$

Ion-pairing will affect all the properties of a solution, and a great many of these—activities (from f.p., v.p., e.m.f. measurements), conductivities, transference numbers, light absorption, optical rotation, reaction rates, distribution and ion-exchange equilibria—have been used in studying it. The extent of ion-pairing varies from an amount too small to measure very accurately in the case of potassium nitrate to values like 75% ion-pairing for manganese(II) oxalate at 0.005 mol $dm^{-3}$. In non-aqueous solutions it is more general and more extensive than in water, and this is due in part to the lower relative permittivity and more powerful electrostatic forces in these solvents.

The *Bjerrum ion-association theory* (q.v.) accounts for the results satisfactorily in some cases, but in others it is clear that additional factors must be taken into account. One of these is hydration—that is, the interaction of the free ions and ion-pairs, respectively, with the surrounding solvent. It has been shown that two types of ion-pair exist in, for instance, copper sulphate solutions:

$$Cu^{2+}, OH_2 + OH_2, SO_4^{2-} \rightleftharpoons (Cu^{2+}, OH_2, SO_4^{2-}) \rightleftharpoons (CuSO_4)$$

Here $OH_2$ represents the shells of polarised water molecules surrounding the ions. In the first reaction the sulphate ion loses some of its closely adhering water and an 'outer sphere ion-pair' is formed; in the second reaction the intervening water is all lost and an 'inner sphere' or 'contact' ion-pair is formed. It is the sum of these two types of ion-pair that is measured by the fall in the molar conductivity or activity, and their relative amounts depend on how much short-range forces between the ions favour the stability of the contact ion-pair. Another 'hydration' effect is found in the salts of the oxy-acids, which tend to show more ion-pairing than other salts of the same valence type. Here there is some evidence that the oxygen atoms of the nitrate ion, for example, readily take the place of water molecules in the hydration shell of the cation, so that contact ion-pairs are more readily formed. Finally, another factor is illustrated by the dissociation constants of the amino-acetates, which vary by more than a millionfold in passing from calcium to copper. Here, as with manganese(II) oxalate, already quoted, the influence of chelation and other structural factors becomes obvious.

*See also* D; and Rosseinsky, D. R. (1971), 'Interactions Involving Aquo-ions', *Annual Reports of the Chemical Society* (*London*), **A68,** 82.

**Ion-selective electrode**

The first ion-selective electrode was the hydrogen-responsive *glass electrode* (q.v.). No electrode is entirely specific towards a particular ion; the presence of other ions can seriously interfere with electrode performance. Thus, electrodes may show a mixed response; the normal glass electrode at $pH > 9$ shows errors in solutions containing sodium ions, attributed to sodium ion response. Variation of the composition of the glass permits this sodium ion response to be extended to lower pH values, eventually showing interference to the hydrogen ion at low pH values. For such electrodes the empirical equation is

$$E = \text{constant} + (RT/F) \ln (a_i + K_{ij}a_j)$$

where $i$ and $j$ refer to two singly charged ions and $K_{ij}$ is the selectivity constant. For response to ion $i$ only, $K_{ij}$ must be small.

Dissolution of active material, poisoning, etc., can also interfere seriously with the functioning of such electrodes.

Ion-selective electrodes may be classified as follows.

(1) Solid state electrodes, which are based on a single crystal or compacted disc of an insoluble salt (e.g. AgX) sealed in a tube (figure I.1). The principal advantage of these over the conventional silver–silver chloride type of electrode is that the presence of oxidising agents in the solution does not affect the e.m.f. A crystal of lanthanum fluoride sealed in a tube permits the direct determination of $F^-$ in solution over the range $10^{-1}$–$10^{-6}$ mol dm$^{-3}$. This electrode is highly reproducible with a response time of only a few seconds; the hydroxyl ion is the only seriously interfering ion.

(2) Heterogeneous membrane electrodes, in which the active material is dispersed in an inert binder to give suitable mechanical properties (figure I.2). Polyvinyl chloride and polystyrene, originally used as inert

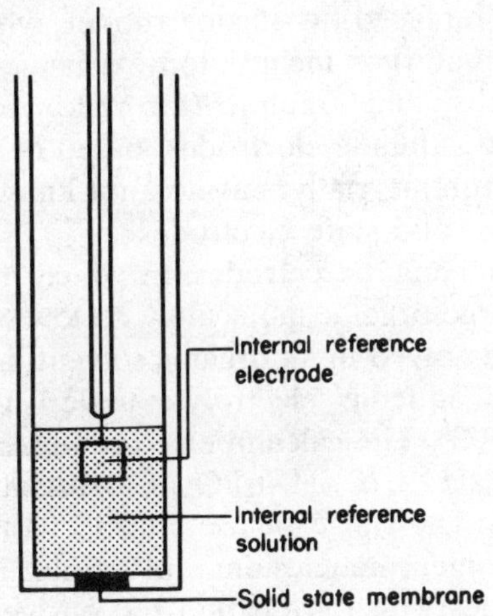


Figure I.1 Solid state electrode

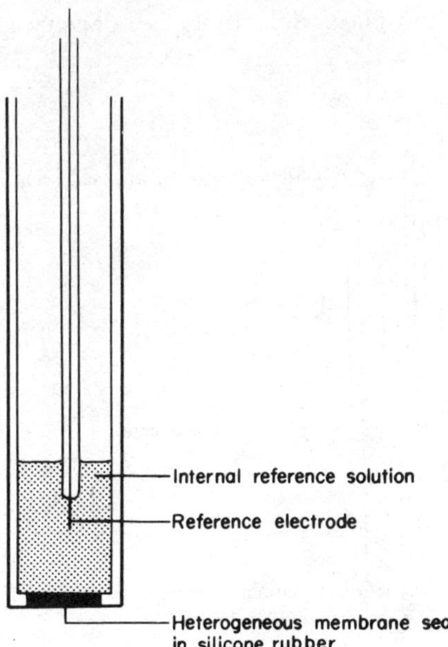


Figure I.2 Heterogeneous membrane electrode

binders, have been replaced by silicone rubber, which has far superior properties. Active materials include ion-exchange resins, precipitates of simple metal salts, and graphite (for redox reactions). Chloride, bromide, iodide and sulphide electrodes, based on the insoluble silver salts, are available commercially, but it is not known if these have any advantages over the solid state electrodes.

(3) Liquid ion-exchanger electrodes, in which the ion of interest is incorporated in a large organic molecule with low water solubility. The organic molecule, dissolved in an organic solvent, is most conveniently separated from the aqueous electrolyte under study by a cellulose acetate film (figure I.3). The calcium electrode uses the calcium salt of didecyl-phosphoric acid, $[(C_{10}H_{21})_2PO_2]_2Ca$, dissolved in di-n-acetyl-phenylphosphonate. The range of usefulness of this electrode is 1 to $10^{-4}$ mol dm$^{-3}$; hydrogen, magnesium and barium ions interfere.

(4) Glass electrodes which are cation-responsive only are similar in design to the conventional hydrogen-responsive glass electrode. Electrodes responsive to $Na^+$, $K^+$, $NH_4^+$, and $Ag^+$ are obtained by varying the composition of the glass. Selectivity is not very good when metal ions are present.

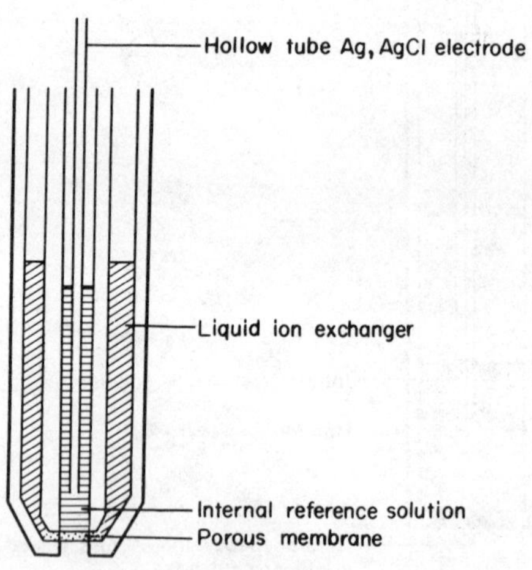


Figure I.3 Schematic diagram of calcium ion-exchange membrane electrode

Table I.1. Ion-selective electrodes

| Solid state | Heterogeneous | Liquid ion-exchanger | Glass |
|---|---|---|---|
| $F^-(LaF_3)$ | $F^-(LaF_3)$ | $Cl^-$ | $H^+$ |
| $Cl^-(AgCl)$ | $Cl^-(AgCl)$ | $ClO_4^-$ | $Na^+$ |
| $Br^-(AgBr)$ | $Br^-(AgBr)$ | $NO_3^-$ | $K^+$ |
| $I^-(AgI)$ | $I^-(AgI)$ | $Ca^{2+}$ | $NH_4^+$ |
| $S^{2-}(Ag_2S)$ | $S^{2-}(Ag_2S)$ | $Cu^{2+}$ | $Ag^+$ |
| $Ag^+(AgCl, AgBr, AgI)$ | $Ag^+(AgCl, AgBr, AgI)$ | $Pb^{2+}$ | $Li^+$ |
| $Cu^{2+}(Ag_2S+CuS)$ | $SO_4^{2-}(BaSO_4)$ | $BF_4^-$ | $Ca^{2+}$ |
| $Pb^{2+}(Ag_2S+PbS)$ | $PO_4^{3-}(BiPO_4)$ | | |
| $Cd^{2+}(Ag_2S+CdS)$ | | | |
| $CN^-(AgI, Ag_2S)$ | | | |

Ion-selective electrodes find many applications, e.g. titrations (EDTA, halide, mixed halides, sulphate), water-hardness determination and treatment, analysis and control of plating baths, and $SO_2$ determination; but care in the selection and use of such electrodes in the presence of possible interfering ions is essential.

*See also* Durst, R. A., Editor (1969), *Ion Selective Electrodes*, N.B.S. Special Publication No. 314, Washington; and Pungor, E., Editor (1973), *Ion Selective Electrodes*, Symp. Proc., Akad. Kiado, Budapest.

# K

**Kohlrausch's law**
*See* Conductivity at infinite dilution.

# L

**Lead accumulator**
*See* Accumulator.

**Limiting current density**

In any electrolytic process ions are used up or are generated at the electrodes, and concentration changes occur at the electrode surfaces, leading to *concentration overpotential* (q.v.). The concentration change near the electrode sets up a concentration gradient, and diffusion of electrolyte into the depleted surface layer (or from a more concentrated surface layer) supplements ionic migration in providing conditions in which electrolysis can continue (cf. *transport number*).

A case of theoretical and historical interest was worked out by H. J. S. Sand in 1901, in which the cell was designed to minimise convective mixing, and diffusion was calculated according to Fick's laws. In normal practice, however, stirring or convection will keep the concentration uniform through the bulk of the solution, and it is assumed that the concentration change is confined to a diffusion layer of thickness $\delta$ in which there is a uniform concentration gradient from the concentration in the bulk of solution to its value at the electrode surface. Calculations based on this assumption have given values of about 0.5 mm for the thickness of the layer in unstirred solutions, down to $\delta = 0.01$ mm for briskly stirred solutions. The assumption is known to be inexact, but theoretical treatment for a rotating disc electrode gives a profile (figure L.1) to which a uniform gradient is a good approximation.

When a constant current is passing through a cell, a steady state is set up in which the ions removed, say, at the cathode by deposition, are exactly replaced by those arriving through electrolytic migration and diffusion. If $m_e$ is the concentration of electrolyte at the electrode surface, and $m$ that of the bulk of solution, the amount arriving by diffusion will be $D(m - m_e)/\delta$ mol cm$^{-2}$ s$^{-1}$, where $D$ is the diffusion coefficient of the electrolyte. The amount arriving by electrolytic migration will be

$$t_c j/nF \text{ g-ion cm}^{-2}\text{ s}^{-1} \qquad \text{(L.1)}$$

where $t_c$ is the cation transport number, $j$ the current density in A cm$^{-2}$ and $nF$ the number of coulombs associated with 1 g-ion (i.e. $n$ is its charge). Together these must make good the loss by discharge, which is $j/nF$ g-ion cm$^{-2}$ s$^{-1}$, so

$$j(1 - t_c)/nF = D(m - m_e)/\delta \qquad \text{(L.2)}$$

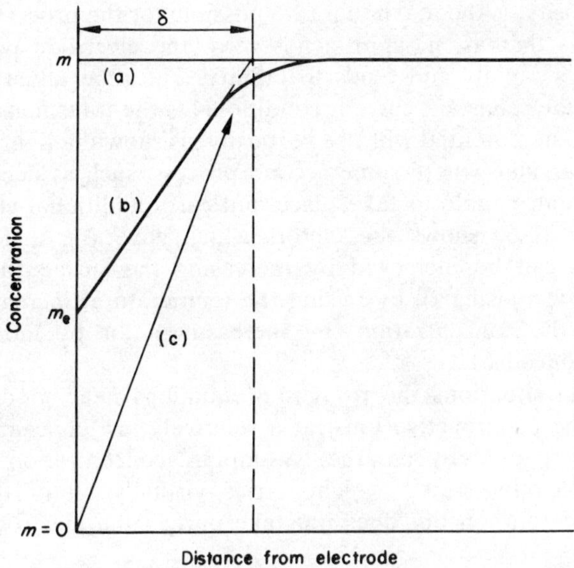


Figure L.1 Diffusion layer: δ is thickness of boundary layer calculated from equation (L.6); (a) initial distribution; (b) theoretical curve for concentration gradient; (c) concentration gradient for $j_{max}$

The same argument would give at a dissolving anode

$$\underset{\text{gain from anode}}{j/nF} = \underset{\text{loss by migration}}{t_c j/nF} + \underset{\text{loss by diffusion}}{D(m_e - m)/\delta} \tag{L.3}$$

The electrode potential corresponding to this steady state is fixed (neglecting any other complications) by the concentration overpotential, which is

$$\Delta E = (RT/nF)\ln(m/m_e) \tag{L.4}$$

If now the applied e.m.f. is increased, there is a limit beyond which the current cannot rise: this is reached when $m_e$ at the electrode surface has fallen to virtually zero. Equation (L.2) now becomes

$$j_{max}(1 - t_c)/nF = Dm/\delta \tag{L.5}$$

Diffusion cannot supply more material than this, and $j_{max}$, the limiting current density, is the maximum rate possible for the process. Equation (L.4) shows that as $m_e$ approaches zero, the electrode potential increases very rapidly and tends to infinity. The time taken for this to happen (under constant current conditions) is the *transition time* (q.v.). In practice the potential will rise vertically, as shown in figure L.2, until it reaches a value where some second process, such as decomposition of the solvent, is able to take place and carry additional current.

Equation (L.5) shows the factors which govern $j_{max}$. The rate of electrolysis can be increased by increasing the surface area of the electrode (decreasing $j$), by raising the temperature (increased $D$), by increasing the concentration, by increasing $t_c$ or by more efficient stirring (reduced $\delta$).

In certain situations the foregoing equations need modifying. For instance, the electroactive ions, at a relatively low concentration, are often accompanied by a large 'swamping' concentration of an *indifferent electrolyte* (q.v.) which carries virtually all of the current through the solution but does not take part in the electrode process.

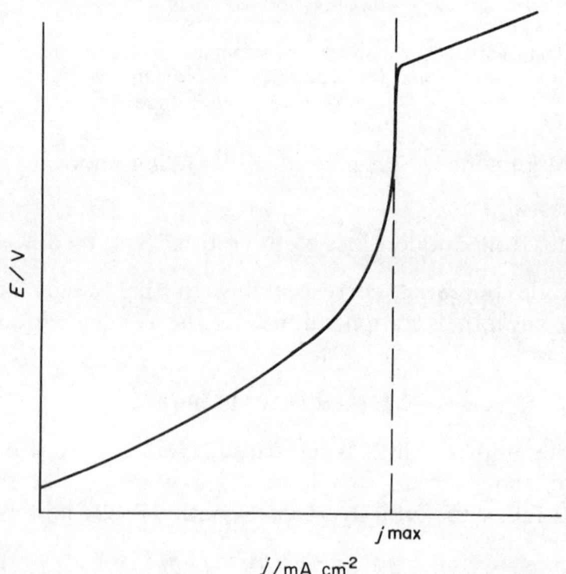


Figure L.2 Potential–current density curve resulting from concentration overpotential

The electrolytic transport term of equation (L.1) is then negligible, and equation (L.2) becomes

$$j/nF = D(m - m_e)/\delta \tag{L.6}$$

and equation (L.5) becomes

$$j_{max}/nF = Dm/\delta \tag{L.7}$$

These equations will also apply to a process such as the cathodic reduction of a non-electrolyte in which electrolytic transport plays no part.

Cases can also arise where diffusion of electrolyte has to be responsible not only for the whole of the material converted at the electrode, but also for sustaining electrolytic migration as well. If, for instance, an anion is being further reduced at a cathode, electrolytic transport will be carrying the reactive ion away from the electrode, and the treatment must start from an equation similar to (L.3):

gain from diffusion = loss by migration + loss by electrolytic action

Finally, another useful relationship may be obtained by combining equation (L.2) with equation (L.5), or equation (L.6) with equation (L.7):

$$j/j_{max} = (m - m_e)/m$$

or

$$m_e/m = (j_{max} - j)/j_{max}$$

**Liquid junction potential**

The liquid junction potential is the potential difference established across the interface between two dissimilar electrolyte solutions. The potential, due to the diffusion of ions across the interface, acts to retard the more rapidly diffusing ions and to accelerate the more slowly diffusing ions, whether they are cations or anions. In this way equilibrium is soon established and a steady liquid junction potential is set up; the magnitude ($<0.1$ V) depends on the *transport number* (q.v.) of the ions, the charge they carry and the concentration of the electrolyte.

The e.m.f. of the *concentration cell* (q.v.),

$$\begin{array}{c|c:c|c} \text{Pt, } H_2(g) & HCl & HCl & H_2(g)\text{, Pt} \\ (101\,325\ \text{N m}^{-2}) & a_2 & a_1 & (101\,325\ \text{N m}^{-2}) \end{array}$$

given by

$$E(\text{cell}) = \frac{2t_- RT}{F} \ln \frac{(a_\pm)_1}{(a_\pm)_2}$$

is the sum of two electrode potentials and the liquid junction potential, $E_j$. The algebraic sum of the two electrode potentials is, theoretically,

$$E_1 - E_2 = \frac{RT}{F} \ln \frac{a(\text{H}^+)_1}{a(\text{H}^+)_2}$$

Hence,

$$E_j = E(\text{cell}) - (E_1 - E_2)$$

$$= \frac{2t_- RT}{F} \ln \frac{(a_\pm)_1}{(a_\pm)_2} - \frac{RT}{F} \ln \frac{a(\text{H}^+)_1}{a(\text{H}^+)_2}$$

Assuming the ratio $a(\text{H}^+)_1/a(\text{H}^+)_2 = (a_\pm)_1/(a_\pm)_2$, then

$$E_j = \frac{(2t_- - 1)RT}{F} \ln \frac{(a_\pm)_1}{(a_\pm)_2} = \frac{(t_- - t_+)RT}{F} \ln \frac{(a_\pm)_1}{(a_\pm)_2}$$

or, in general, for cation-reversible electrodes,

$$E_j = \left(t_- \frac{\nu}{\nu_+} - 1\right) \frac{RT}{nF} \ln \frac{(a_\pm)_1}{(a_\pm)_2}$$

and for anion-reversible electrodes,

$$E_j = \left(t_+ \frac{\nu}{\nu_-} - 1\right) \frac{RT}{nF} \ln \frac{(a_\pm)_1}{(a_\pm)_2}$$

For cation-reversible electrodes, when $t_+ = t_-$, $E_j = 0$; when $t_- > t_+$, $E_j$ is positive and adds to the sum of the electrode potentials; and when $t_+ > t_-$, $E_j$ is negative and yields a lower e.m.f. for the cell than the sum of the electrode potentials. Attempts have been made to eliminate liquid junction potentials by interposing a salt bridge consisting of a concentrated solution of KCl, $KNO_3$, etc., for which $t_+ \approx t_-$. As the solution is so concentrated, most of the diffusion is due to the bridge electrolyte. Thus the junction potential of the original cell is replaced by two junction potentials which are acting in opposition and which have values very near zero. Junction potentials may be reduced by this method, but it is doubtful if they are completely eliminated.

*See also* G.

**Lithium, electrometallurgy**

Lithium is produced by an electrolytic process broadly similar to that used for sodium (see *sodium, electrometallurgy*). The electrolyte is a fused mixture of lithium and potassium chlorides at about 680 K. Lithium is preferentially liberated at the cathode; its standard potential in water [$E^{\ominus}(Li^+, Li) = -3.05$ V] is more negative than that of potassium [$E^{\ominus}(K^+, K) = -2.93$ V], but in the absence of water the order is reversed. Moreover, the conductivity of LiCl is nearly four times that of KCl, and the complexing of the latter in a mixture would result in a much higher activity of the lithium ion. The electrolysis is conducted at about 8 V, and the current efficiency is 85–90%.

**Luggin capillary**

*See* Decomposition voltage.

# M

**Magnesium, electrometallurgy**

Magnesium is obtained by electrolysing a fused mixture of magnesium, sodium and calcium chlorides in the approximate proportions 25:60:15. This mixture has a density greater than that of molten magnesium and the cell is operated at a temperature somewhat above the m.p. of magnesium (924 K), so that the molten metal floats and is removed from the top of the cell. The steel container acts as cathode, and a central graphite anode is contained in a porous pot. The chlorine that is liberated rises and is piped away for re-use in making the chloride.

The process is straightforward if pure anhydrous materials are used; the current efficiency is then over 90% and the product very pure. The main complication in the process, on account of the deliquescence of the starting materials, is contamination by water. Not only will any water present be electrolysed first, causing a drop in current efficiency, but it will also create magnesium oxide which sinks to the bottom taking some magnesium with it, thus causing an additional loss.

**Membrane electrode**
*See* Ion-selective electrode.

**Membrane potential**
*See* Donnan membrane equilibrium‡

**Mercury cell**
The over-all reaction of the mercury dry cell is

$$Zn + HgO \rightarrow Hg + ZnO$$

The electrolyte is a concentrated potassium hydroxide solution, and the zinc ions dissolving from the zinc anode are precipitated as zinc hydroxide and oxide. The cathode is a compressed mixture of mercury(II) oxide and graphite, and the cathode reaction is

$$HgO + H_2O + 2e \rightarrow Hg + 2OH^-$$

The electrolyte is not consumed in the reaction, so that little is necessary. The capacity–volume ratio of the cell is therefore high, and it can be made in very small sizes for deaf aids, etc. The cell has other advantages: the voltage (1.35 V) is very stable, and it can give high currents without loss in performance; its storage life is also high. In some miniature cells, cadmium replaces zinc as anode.

**Metal electrode**
A metal electrode is reversible to the metal ion in solution:

$$M \rightleftharpoons M^{n+} + ne$$

The *electrode potential* (q.v.) is given by

$$E(M^{n+}, M) = E^{\ominus}(M^{n+}, M) + \frac{RT}{nF} \ln a(M^{+})$$

All the metal electrodes can be made by soldering or otherwise attaching a short length of wire or rod onto a piece of copper wire, sealed into a piece of glass tubing so that the metal protrudes through the end of the glass. No solution must come into contact with the copper–metal junction. They all have low electrical resistance and can be used with a simple potentiometer circuit.

**Metal–metal oxide electrode**
*See* Antimony–antimony oxide electrode.

**Mobility**
*See* Molar ionic conductivity.

**Molar ionic conductivity**
Recent recommendations on units and symbols define the mobility of an ion, $u$, as its velocity under unit potential gradient. Most writers in the past, however, have used the term as a convenient synonym for the equivalent ionic conductivity, $\Lambda_i$. The relation between the two is $\Lambda_i = Fu_i$, and this should be remembered when consulting the literature.

The ionic conductivity is obtained by combining the conductivity with the *transport number* (q.v.). The cation transport number of an electrolyte measures the fraction of the current carried by the cation, so the product of this and the molar conductivity of a completely dissociated binary electrolyte gives the molar conductivity of the cation: $t_c\Lambda = \Lambda_c$, and similarly $t_a\Lambda = \Lambda_a$.

Very accurate values are available for the common ions in water at 298 K, based on the work of MacInnes, Shedlovsky and Longsworth. These authors measured transport numbers and conductivities of HCl, KCl, NaCl and LiCl over a range of concentration, thus obtaining values for the molar conductivity of the chloride ion in these solutions. These were extrapolated according to Shedlovsky's equation (see *conductivity at infinite dilution*) with the result shown in figure M.1. A very accurate value, $\Lambda^{\infty}(Cl^-) = 76.34\ \Omega^{-1}\ cm^2\ mol^{-1}$, is thus available for the chloride ion, and by subtraction of this from the $\Lambda^{\infty}$ values of the four electrolytes, $\Lambda_i^{\infty}$ values for the four cations are obtained. Other transport number data are now unnecessary, as the molar conductivity of a cation can be found by subtraction from the $\Lambda^{\infty}$ value of its chloride, or of an anion from $\Lambda^{\infty}$ for its sodium or potassium salt. Values for the commoner ions are given in table M.1.

The $H^+$ and $OH^-$ ions have relatively very high values, and this is due to the strong tendency to hydrogen bonding between water molecules. As a result, the proton is able to move in the direction of the current by an exchange of partners much more readily than it could

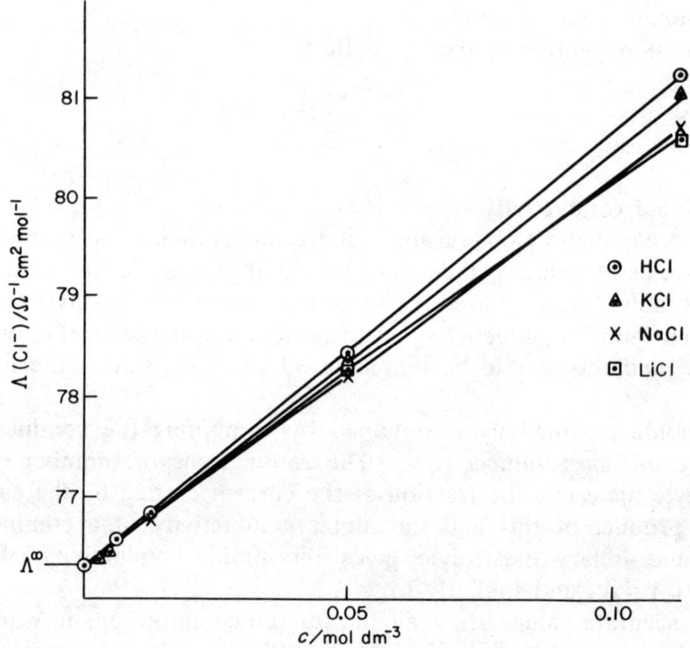


Figure M.1 Ionic conductivity of chloride ion

Table M.1. Molar ionic conductivity/$\Omega^{-1}$ cm$^2$ mol$^{-1}$ at 298 K

| Cations | $\Lambda_i^\infty$ | Anions | $\Lambda_i^\infty$ |
|---|---|---|---|
| $H^+$ | $349.8_1$ | $OH^-$ | 198.3 |
| $Li^+$ | $38.6_8$ | $F^-$ | 55.4 |
| $Na^+$ | $50.1_0$ | $Cl^-$ | $76.3_4$ |
| $K^+$ | $73.5_0$ | $Br^-$ | $78.1_4$ |
| $Ag^+$ | $61.9_0$ | $I^-$ | $76.8_4$ |
| $NH_4^+$ | $73.5_5$ | $NO_3^-$ | $71.4_6$ |
| $\frac{1}{2}Mg^{2+}$ | $53.0_5$ | Acetate | $40.9_0$ |
| $\frac{1}{2}Ca^{2+}$ | $59.5_0$ | Picrate | 31.4 |
| $\frac{1}{2}Cu^{2+}$ | 56.6 | $\frac{1}{2}SO_4^{2-}$ | $80.0_2$ |
| $\frac{1}{3}La^{3+}$ | $69.7_5$ | $\frac{1}{3}Fe(CN)_6^{3-}$ | 100.9 |

by a normal process of migration. The proton-jump mechanism for water is

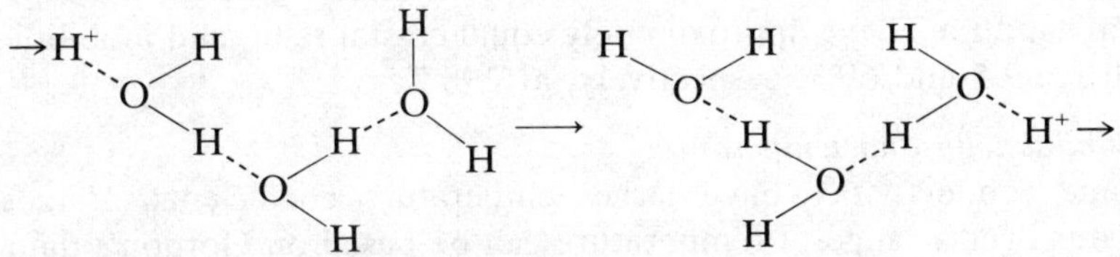

```
                       H
→H⁺     H              |         H     H       H
   ·.  /               O    ⟶     \   /         \
     O              ·'  \          O ·.          O ·.
      \            H     H              H      H      H⁺→
       H ·.      /                       ·.  /
           O                               O
            \                             /
             H                           H
```

and a similar mechanism will apply to the $OH^-$ ion. The other ions are more nearly similar in their values. A low figure for the picrate ion reflects its large size, and the same influence is seen in homologous series of organic acid ions, where mobilities fall regularly with increasing size. For the alkali metals, however, the values are in inverse order of size. This is attributed to 'hydration' effects; the small lithium ion orients and holds water molecules on account of the strong ion–dipole attraction, whereas the weaker attraction of the larger ions leads to a net 'structure-breaking effect' through its disruptive influence on the hydrogen bonding of the water molecules in its vicinity.

*Valency effects*
The molar conductivity of an ion depends on the charge it carries, and on its mobility under unit potential gradient. In making comparisons the first effect is usually eliminated; thus, the molar conductivity of the sulphate ion at 298 K might be quoted as: $\Lambda^\infty(SO_4^{2-}) = 160.04$, but is more usually given as $\Lambda^\infty(\frac{1}{2}SO_4^{2-}) = 80.02$. The values for other multivalent ions in table M.1 are given in the same way. On this basis we are now left with the second effect: the force on an ion moving in unit field is proportional to its charge, so that (other things being equal) a proportionality between mobility and valency might be expected. This is almost realised among cobaltammine ions:

$$\Lambda^\infty([Co(NH_3)_4(NO_2)_2]^+) = 36.3$$
$$\Lambda^\infty(\tfrac{1}{2}[Co(NH_3)_5Cl]^{2+}) = 70.4$$
$$\Lambda^\infty(\tfrac{1}{3}[Co(NH_3)_6]^{3+}) = 101.9$$

With smaller ions, however, the positive effect of increasing charge is partly offset by the stronger ion–solvent interactions. For dibasic acids it is a rough rule that the mobility of the $X^{2-}$ is about 1.7 times that of the $HX^-$ ion, and this tendency for mobility to increase less rapidly than valency reaches its extreme for the simple elementary ions. $Na^+$, $Ca^{2+}$ and $La^{3+}$ have approximately equal crystal radii, and mobilities of 50.1, 59.5 and 69.7, respectively, at 298 K.

*Conductivity and temperature*
Ionic conductivities have large temperature coefficients. Values in water over a range of temperature can be based on Gordon's data for the chloride ion (table M.2).

Table M.2

| Temperature/K | $\Lambda^\infty(Cl^-)/\Omega^{-1}\ cm^2\ mol^{-1}$ |
|---|---|
| 288 | 61.42 |
| 298 | 76.35 |
| 308 | 92.21 |
| 318 | 108.90 |

From these values those of the other ions can be obtained. A survey then shows that the main influence is the changing viscosity of the solvent. If Stokes's law were exactly obeyed, the product $\Lambda_i^\infty \eta$ would be constant. This is approximately true of large ions such as picrate and $NEt_4^+$; it is also true of heavily 'hydrated' ions such as $Li^+$ and $La^{3+}$, suggesting that the 'effective radius' of these ions is unchanged by temperature. For the larger univalent ions such as $K^+$, $Cl^-$, $Br^-$, the product falls with increasing temperature, by about 30% between 237 K and 373 K. This presumably reflects a reduced 'structure-breaking' effect as the hydrogen bonding of the solvent diminishes, and is in line with other evidence on the point.

*Ionic conductivities in non-aqueous solvents*
Data here are much more fragmentary owing to the shortage of measured transport numbers. In their absence approximate values can be obtained from 'Walden's rule': $\Lambda_i^\infty \eta = \text{constant}$. Walden's data supported the constancy of this Stokes's law product for tetraethylammonium picrate in a large number of solvents, and approximate values for other ions may be based on the value: $\Lambda_i^\infty \eta = 0.270\ \Omega^{-1}\ cm^2\ mol^{-1}$ poise for the picrate ion, which is the known value in water.

*See also* R & S.

**Moving boundary method**

The moving boundary method of determining a *transport number* (q.v.) depends on observing the movement, under a constant current, of the boundary between the solution under study and a following 'indicator' solution. For accuracy this boundary must remain quite sharp, and the conditions for this are as follows: (a) the two electrolytes have a common anion when the movement of a cation is being observed, and vice versa; (b) the conductivity of the indicator ion must be lower than that of the leading ion; (c) the solution with the lower density must be above the other; (d) an initially sharp boundary must be established; and (e) the concentrations of the two solutions must be in the correct ratio.

The necessary concentration ratio can be deduced from figure M.2, in which the electrolyte being studied, MA, is initially separated from the following electrolyte, NA, by a sharp boundary at pq. While $x$ Faradays of electricity are passed through the solution, the boundary moves to rs, and must remain perfectly sharp. In the volume V,

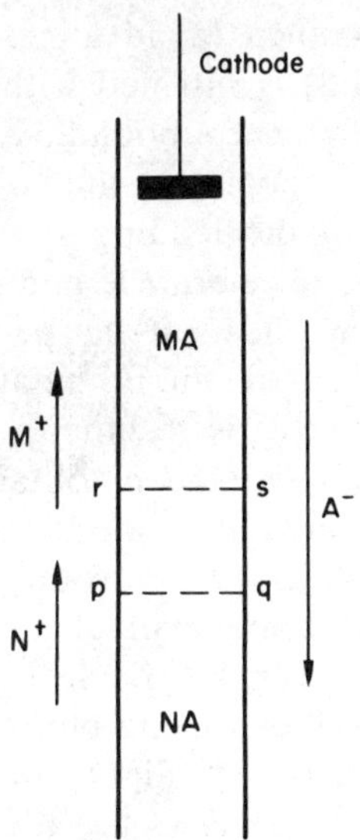


Figure M.2 Moving boundary method for transport number determination; condition for sharp boundary

enclosed by pq and rs, MA has therefore been entirely replaced by NA. The quantity of $M^+$ transported across rs by the current is $t_{M^+}x$ mole, and, as this is the amount that was previously contained in the volume $V$,

$$t_{M^+}x = c_1 V \tag{M.1}$$

where $c_1$ is the concentration of MA.

Similarly the quantity of $N^+$ transported across pq is $t_{N^+}x$ mole and again

$$t_{N^+}x = c_2 V \tag{M.2}$$

where $c_2$ is the concentration of NA. Dividing equation (M.2) by equation (M.1) gives the required relation:

$$c_2/c_1 = t_{N^+}/t_{M^+} \tag{M.3}$$

The indicator ion, with the lower conductivity and transport number, must also be at a correspondingly lower concentration.

To satisfy this condition exactly, the transport number being measured would have to be known in advance. Fortunately it has been found that if equation (M.3) is satisfied within 10%, there will be an automatic adjustment to a sharp boundary. The reason is that the potential gradient will be steeper in the following solution, with its lower conductance, than in the leading solution; hence, any $M^+$ ion behind the front would be accelerated, and any $N^+$ ion ahead of the boundary would slow down. This self-sharpening tendency can also be seen if the current is interrupted; the boundary then loses its sharpness but regains it when the current is resumed.

Very accurate results have been obtained by MacInnes and Longsworth with the apparatus illustrated in figure M.3. The electrode vessels end in glass discs which can be rotated into or out of line with the graduated tube B. The upper pair of discs is used for descending boundaries and the lower pair for rising boundaries. In the former case, B and the cathode vessel will be entirely filled with the leading solution, and with the discs out of alignment the anode vessel is filled with the indicator solution. The discs are then slid into line to form a 'sheared boundary' at C, and the experiment is started. The movement of the boundary can be detected by the difference in refractive index of

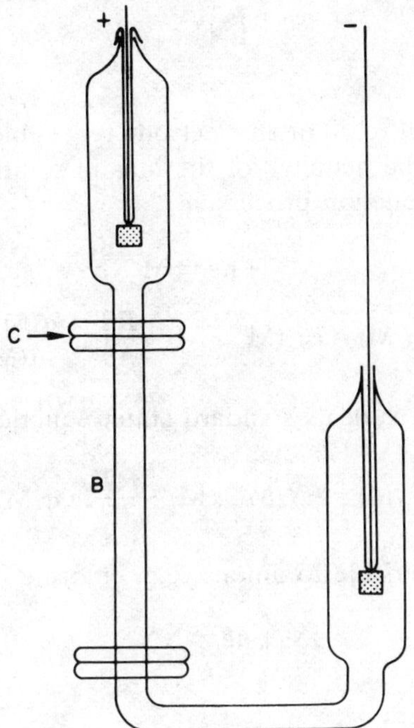


Figure M.3 Moving boundary apparatus

the two solutions, and the transport number is calculated from equation (M.1). If a constant current of $I$ amperes has been passed for $s$ seconds, $x = Is/F$, and

$$t_+ = cVF/Is$$

The volume, $V$, swept out by the boundary is measured from the graduations of tube B, but in accurate work a correction must be made for the small volume changes resulting from the electrolysis.

In a variant of this method the two solutions are originally separated by a porous disc, and the material passing through this during the electrolysis is determined analytically.

*See also* H & O, J & P.

# N

## Nernst equation

The *electrode potential* (q.v.) of an electrode reversible to cations (e.g. a metal) is related to the *activity*‡ of the ions in solution by the Nernst equation. For the reduction process,

$$M^{n+} + ne \rightleftharpoons M$$

$$E(M^{n+}, M) = E^{\ominus}(M^{n+}, M) + \frac{RT}{nF} \ln \frac{a(M^{n+})}{a(M)} \qquad (N.1)$$

Since the solid metal is in its standard state, equation (N.1) becomes

$$E(M^{n+}, M) = E^{\ominus}(M^{n+}, M) + \frac{RT}{nF} \ln a(M^{n+})$$

For an electrode reversible to anions (e.g. chlorine electrode),

$$\tfrac{1}{2}X_2 + ne \rightleftharpoons X^{n-}$$

the Nernst equation is

$$E(X^{n-}, X_2, Pt) = E^{\ominus}(X^{n-}, X_2, Pt) - \frac{RT}{nF} \ln \frac{a(X^{n-})}{a^{1/2}(X_2)} \qquad (N.2)$$

For gases, assuming ideal behaviour the activity can be replaced by the partial pressure and, hence,

$$E(X^{n-}, X_2, Pt) = E^{\ominus}(X^{n-}, X_2, Pt) - \frac{RT}{nF} \ln \frac{a(X^{n-})}{p^{\frac{1}{2}}(X_2)}$$

For a *redox electrode system* (q.v.),

$$\text{Oxidised state} + ne \rightleftharpoons \text{Reduced state}$$

the corresponding equation is

$$E(R, O) = E^{\ominus}(R, O) + \frac{RT}{nF} \ln \frac{a(\text{oxidised state})}{a(\text{reduced state})} \qquad (N.3)$$

**Nickel, electrometallurgy**

A major proportion of the nickel produced is now obtained electrolytically. The concentrate reaches the electrolysis stage as matte, cast in the form of anodes. These consist mainly of the sulphide, $Ni_3S_2$, together with metallic nickel, and copper, iron and cobalt as the most important impurities. In the electrolytic tanks the nickel cathodes are separated from the matte anodes by diaphragms of synthetic cloth. The electrolyte, a solution of nickel sulphate and nickel chloride, enters the cathode compartment, overflows from the anode compartment and is chemically treated before recirculation. Electrolysis requires a cell voltage of 3–4 V. The whole of the sulphur remains in the anode sludge together with some precious metal residues, from which it is separated by melting and filtration.

The current efficiency at the matte anodes is only about 95%, so that a deficiency of nickel and an increase in acidity develop in the electrolyte. The chemical treatment of the latter must therefore correct for this as well as remove impurities. The electrolyte at pH 1.5 is first treated with $H_2S$, which precipitates copper and arsenic(III) sulphides; copper is recovered from the precipitate. The electrolyte then flows to a series of agitated aeration tanks and finely ground anode scrap is added. Metallic nickel dissolves and the sulphide undergoes the reaction

$$Ni_3S_2 + O + 2H^+ \rightarrow Ni^{2+} + 2NiS + H_2O$$

The filtered solution is next treated with chlorine to oxidise the iron and cobalt; these are precipitated at a controlled pH by the addition of nickel carbonate. Cobalt is recovered from the precipitate, and the filtrate is now ready for return to the electrolytic tanks.

**Non-aqueous solutions**

For electrochemical purposes, non-aqueous solvents have the potential advantage of enabling processes to be carried out which are impossible in water. A restriction to their use, however, is the limited solubility of most electrolytes, and a further special problem is the need to remove traces of water (and oxygen) and to work out of contact with moist air. Thus a $10^{-3}$ mol dm$^{-3}$ solution of HCl in a pure non-basic solvent is non-conducting; the addition of only 0.018 g of water per dm$^3$, however, converts it into the strong electrolyte $H_3O^+$, $Cl^-$.

*Non-aqueous solutions*

*Conductances of salts*

The solvent influences the conductance of a salt through three main factors: (a) its viscosity, (b) its dielectric constant and (c) its interaction with the ions (solvation), which will be determined in a complex way by its own structure, its molecular size, and the location and orientation of any dipole(s) it contains.

The molar conductivity at infinite dilution is governed primarily by the viscosity; thus the molar conductivities of the common ions are much higher in acetone than they are in water. Where data are lacking, approximate values for $\Lambda^{\infty}$ can be calculated from those for another solvent by using Walden's equation:

$$\Lambda^{\infty}\eta = \text{constant}$$

This equation ignores factor (c), but is fairly reliable for large ions, such as the tetra-alkylammonium ions, and for the more non-polar solvents, where solvation will not be expected to be critically important.

Measurements at very low concentrations have confirmed the validity of *Onsager's equation* (q.v.) for a large number of systems, and it can be assumed to hold generally. At finite concentrations, however, the dielectric constant of the solvent becomes the most important factor. Values for some of the solvents which have been employed (Table A.II, p. 244) are generally considerably lower than that of water. In these, coulomb attractions are far stronger, the Onsager slopes are much steeper, and ion-pairing is much more general and more extensive than in water, so that the conductance is liable to fall off rapidly to low values.

The straight line (figure N.1) obtained by plotting the p$K$ of tetraethylammonium picrate (circles), measured in a range of solvents, against the reciprocal of the dielectric constant is that which would be expected if ion-pairing was governed purely by electrostatic forces. For these two large ions the picture does not appear to be complicated by solvation effects or by special (non-coulombic) attractive forces between the ions.

Lithium picrate (p$K$ values shown by triangles) presents a different picture, the points being widely scattered. This is attributed to the influence of the third factor mentioned, 'solvation'. Four of the points

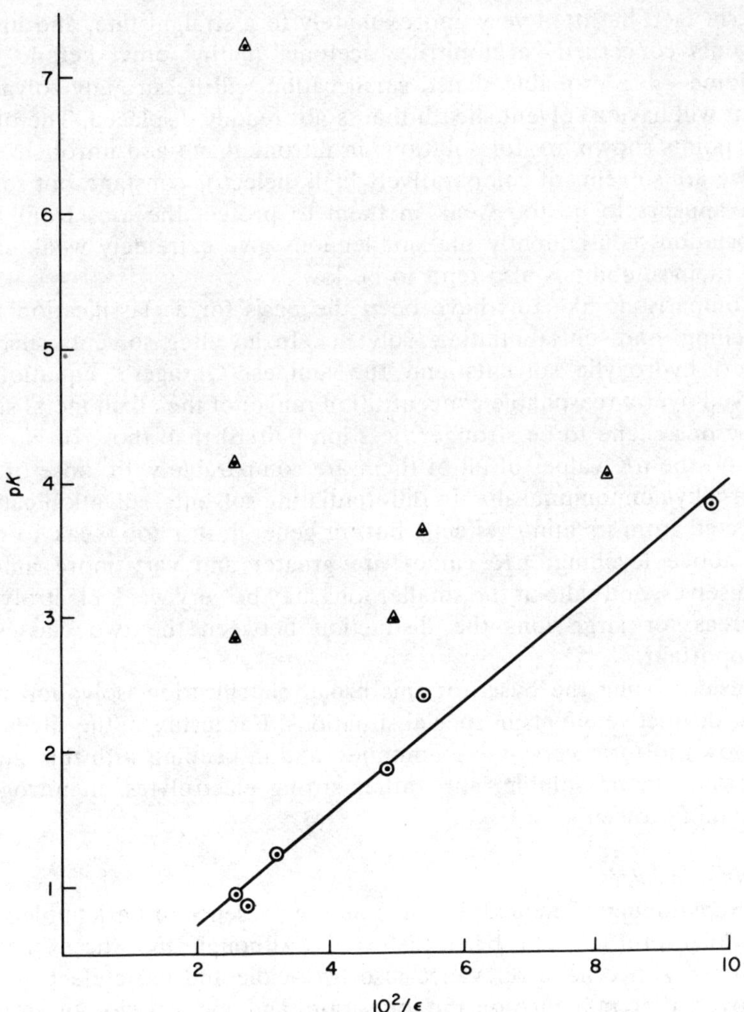


Figure N. 1 p$K$ values in non-aqueous solvents: ⊙ tetraethylammonium picrate; △ lithium picrate

can, in fact, be fitted very approximately to a straight line, and in the solvents concerned—acetonitrile, acetone, methyl ethyl ketone and pyridine—it is probable that a small cation will be strongly solvated, i.e. it will have a solvent sheath that is not readily displaced. The other two points shown are for solutions in nitromethane and nitrobenzene. These are solvents of comparatively high dielectric constant, but solvation appears to be too weak in them to protect the ions from ion-association; consequently the smaller ions give extremely weak salts, and their solubilities also tend to be low.

Comparisons like this have been the basis for a classification into 'levelling' and 'differentiating' solvents. In levelling solvents such as water, hydroxylic solvents and the amines, Onsager's equation is obeyed over a reasonable concentration range; of the alkali metal salts, those of Li tend to be stronger (less ion-paired) than those of Na and K; and the p$K$ values of all of them are comparable with those of the tetra-alkylammonium salts. In differentiating solvents, solvation can be detected from its minor effects, but, in general, it is too weak to give the above levelling; p$K$ values are greater and vary more among themselves, and salts of the smaller ions may be very weak electrolytes, whereas for large ions the distinction between the two classes is unimportant.

Besides being the basis for this broad classification, solvation may show distinctive effects in special situations. For instance, the silver ion is known to form very stable ammines, and in keeping with this, silver salts tend to be soluble, and rather strong electrolytes, in nitrogen-containing solvents.

### *Acids and bases*

The functioning of an acid depends on the presence of basic molecules to which protons can be transferred. Although the three factors discussed above have relevance also for acidic and basic electrolytes, the overriding influence on the ionisation, and conductivity, of an acid will be the basic strength of the solvent, and vice versa. In acetic acid or sulphuric acid, therefore, substances may ionise as bases which are non-electrolytes in water. In the same way, substances whose acid properties are too feeble to be evidenced in aqueous solutions can be studied, and put in relative order of acid strength, by conductance measurements in ammonia, or other basic solvent. Equally, acids which are completely dissociated in water (e.g. $HClO_4$, $HCl$ and $HNO_3$) reveal their relative strengths as acids when studied in a less basic solvent than water, such as acetic acid or acetonitrile.

*Electrolytic processes*
The standard *electrode potential* (q.v.) is often not greatly different in non-aqueous solvents from that in water, although displacements due to differences in the strength of solvation of the ions are to be expected. The same reference electrodes as are used in water are also usually satisfactory. The rates of electrochemical reactions, however, can be radically altered by changes of solvent, since all the factors which govern the ease of transfer of electrons across the electrode surface are likely to be modified. These include the solvation of the electroactive ions, their tendency to ion-pairing and complex formation, the adsorbability of the solvent and of active species at the electrode surface, and the other factors that may affect the structure of the *electrical double layer* (q.v.).

Several of the electroanalytical methods discussed elsewhere have been successfully applied in non-aqueous solvents. The other main interest centres round the electrodeposition of the active metals, Na, Ca, Al, etc., for which mixtures of fused salts are generally used. Solvents such as formamide, acetonitrile or ethylene diamine are possible, but their usefulness is limited by the need for sufficiently well-conducting solutions, and other practical difficulties. Beryllium, titanium and zirconium have been obtained by electrolysing ethereal solutions. The pure salts do not conduct in ether, but by using mixtures, e.g. of $LiBH_4$ and the metal chloride or bromide, the necessary conductivity is achieved and good deposits of the metals have been obtained.

*See also* Conductance minima; D, L.

# O

**Ohm**
The ohm, $\Omega$, is the unit of electrical resistance (dimensions: $\varepsilon^{-1}l^{-1}t$; units: $kg\ m^2\ s^{-3}\ A^{-2}$ or $V\ A^{-1}$) through which a potential difference of 1

volt will produce a current of 1 ampere:

$$E = IR$$

The International ohm is the resistance offered to an unvarying current by a column of mercury at 0 °C, 14.4521 g in mass, of constant cross-sectional area and 106.300 cm in length.

*See also* Electric units.

**Onsager's equation**

*See* Conductance equations.

**Oscillometry**

Oscillometry is a method of following the changes in conductance or in permittivity, $\varepsilon$, which occur during titration, or during the mixing of two liquids of different $\varepsilon$. The conductance 'cell' is a beaker or test-tube, connected into an oscillatory circuit operating within the frequency range 1–400 MHz. The connection may be through bands of conducting material (e.g. copper foil) wrapped around the outside of the tube and wired into the oscillator, or the cell may be placed inside the oscillator inductance (the 'tank coil'). In either case, changes within the cell alter the loading upon the oscillatory circuit, and changes in its

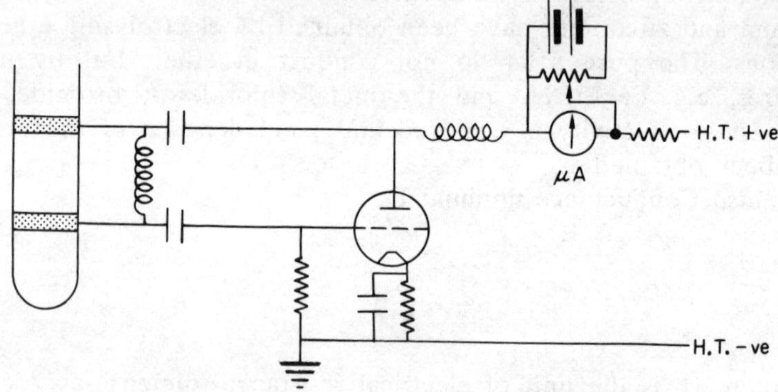


Figure O.1 Simple oscillometer circuit

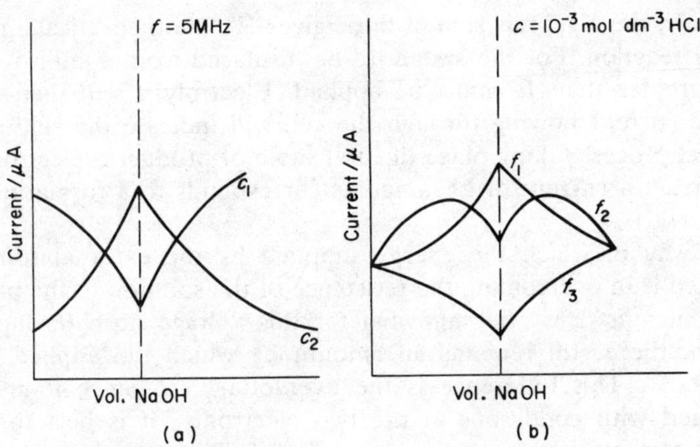


Figure O.2 Typical titration curves: (a) $c_1$ 0.02 mol dm$^{-3}$ HCl against 0.2 mol dm$^{-3}$ NaOH; $c_2$ 0.002 mol dm$^{-3}$ HCl against 0.02 mol dm$^{-3}$ NaOH. (b) $f_1 = 5$ MHz: $f_2 = 10$ MHz; $f_3 = 20$ MHz

frequency, or in the anode or grid current of the oscillator, are observed.

The high frequency used is necessary so that changes can be observed through the glass or silica wall of the cell, via an effective capacitor having this wall as dielectric. A typical circuit and examples of titration curves are shown in figures O.1 and O.2. The frequency or current measurements are related to comparative conductances only, but the technique has been used in a wide variety of applications.

*See also* Pu; and Lee, W. H. (1969), in *Electrometric Methods*, Ed. Browning, D. R., McGraw Hill.

**Ostwald's dilution law**
*See* Arrhenius electrolytic dissociation theory.

**Overpotential**
*See* Activation overpotential; concentration overpotential; Overvoltage.

**Overvoltage**
When an electrolytic cell is under equilibrium conditions no current passes through the cell, and both electrodes are at their equilibrium

potentials; the algebraic sum of these gives $E^\circ$, the theoretical e.m.f. of the cell reaction. For the system to be displaced from equilibrium, an e.m.f. greater than $E^\circ$ must be applied. Electrolysis will then begin, and the current flowing through the cell will indicate the rate of the chemical process taking place (if for 1 mole of product $z$ electrons are transferred, a current of $zF$ amperes corresponds to a formation rate of 1 mol $s^{-1}$).

One way in which the energy supplied by the external source is expended is in overcoming the resistance of the solution to the passage of the current. But after allowing for this voltage drop through the solution, there still remains an amount by which the applied e.m.f. exceeds $E^\circ$. This difference is the overvoltage, $\eta$. Since it must be concerned with conditions at the two electrodes, it is best to study these separately (see *decomposition voltage*). The extent by which the cathode potential is found to be more negative than its equilibrium value is the cathode overpotential, and, similarly, a working anode will be more positive than its equilibrium value by an amount corresponding to the anode overpotential.

At least two, and sometimes three, causes of overpotential can be distinguished. In the first place, the passage of current immediately alters the concentration of the reactive ion at the electrode surface, and the potential needed to maintain current flow will increase. This is called *concentration overpotential* (q.v.) or concentration polarisation. A second type of overpotential is *activation overpotential* (q.v.). The relative importance of this varies greatly; for a process that takes place readily it will be small, but it can be large for an electrode process that requires a high activation energy. The third cause of overpotential is present when an electrode is covered by an adherent surface layer of poorly conducting material. The resistance may be so high that a very large potential drop through this surface film is included in the overpotential measurement.

The overpotential at an electrode (for a given current density) is the sum of these three effects: the concentration, activation and film-resistance overpotentials:

$$\eta = \eta_{\mathrm{C}} + \eta_{\mathrm{A}} + \eta_{\mathrm{R}}$$

The third is really a spurious effect which is easily recognised, if present. Of the other two, the first is concerned with the solution side of the interface, the second with the electron transfer over the interface. They can be distinguished by the fact that the second grows very rapidly when the polarising current is switched on and decays equally rapidly when it is interrupted, while, with the first, these changes are relatively slow, since they depend on diffusion. Stirring greatly affects $\eta_c$ for the same reason.

**Oxidation–reduction systems**
*See* Redox electrode system.

**Oxygen electrode**
The oxygen electrode consists essentially of a lead anode and a silver cathode (figure O.3). The lead anode, covered with a porous polythene

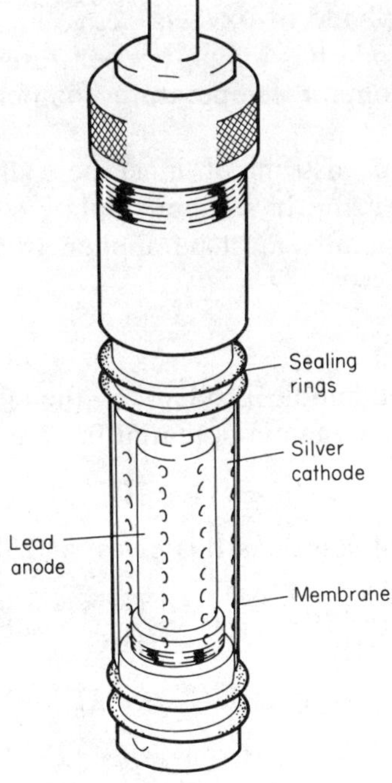


Figure O.3 Oxygen sensor

membrane, is surrounded by the cathode. The silver cathode is a hollow perforated cylinder covered with a polythene membrane which is permeable to oxygen but which is impermeable to water and interfering ions. The electrode core is filled with a solution of $KHCO_3$ and $Na_2CO_3$. No external polarising voltage is necessary; the electrode relies on the diffusion of oxygen through the gas-permeable membrane and the subsequent reduction of oxygen at the cathode to give a current which is proportional to the partial pressure of oxygen in the sample (in the range 0–200% saturation). The electrode reactions are:

at the anode: $Pb \rightarrow Pb^{2+} + 2e$

at the cathode: $O_2 + 2H_2O + 4e \rightarrow 4OH^-$

The aqueous solution flows past the electrode and the dissolved oxygen diffuses through the membrane. Minimum flow conditions are critical since, if the flow rate is too low, the sample water around the electrode will be depleted of oxygen.

The oxygen electrode has a high temperature coefficient (*c.* 6% per degree change); automatic temperature compensation is provided in modern instruments.

The electrode and instrument must be calibrated before use in air-equilibrated water and in sodium sulphite solution.

A correction for salinity must be applied when making determinations on sea-water.

**Oxygen overpotential**

Oxygen is the usual anodic product of the electrolysis of aqueous solutions. The thermodynamic potential for the reaction

$$2H_2O \rightarrow O_2 + 4H^+ + 4e$$

is 1.229 V. In neutral solutions this gives, at 298 K,

$$E = 1.229 + \frac{RT}{4F} \ln \frac{a^4(H^+)}{a^2(H_2O)} = 0.815 \text{ V}$$

and in alkaline solutions, $a(OH^-) = 1$, $a(H^+) = 1 \times 10^{-14}$, it becomes

$$4OH^- \rightarrow O_2 + 2H_2O + 4e, \qquad E = 0.401 \text{ V}$$

These are theoretical values. A reversible oxygen electrode cannot be realised in practice, and at finite currents the evolution of oxygen only occurs at considerable overpotentials on all metals. In contrast with hydrogen evolution, the overpotential is highest on platinum and gold, and relatively low on cobalt, iron and nickel.

In spite of their importance, the oxygen evolution and reduction reactions have had less study than has hydrogen evolution. They present greater difficulties than the latter, partly so because so many metals will dissolve anodically before the potential for oxygen evolution is reached. Also, although reliable Tafel slopes have been obtained in some very careful work, there are indications in other cases of changes with time, which may be associated with changes in the electrode surface. Oxygen is strongly adsorbed, and the result may be a monolayer of adsorbed oxygen, or an oxide layer several molecules thick. A further complication in mechanism studies is the number of intermediates, e.g. peroxide species, which may conceivably participate in the reaction path.

A simple mechanism, for which there is some good evidence, is in the reduction of oxygen at an iridium electrode. The reaction path appears to be

$$O_2 + 2M \rightleftharpoons 2MO$$
$$MO + H^+ + e \rightarrow MOH + H_2O$$
$$MOH + H^+ + e \rightarrow M + H_2O$$

MO is here a surface oxide, and the second is the rate determining step.

*See also* B & R.

# P

**Passivity**

A metal surface is said to be passive when, although exposed to conditions in which it is thermodynamically unstable, it remains unattacked indefinitely. Thus, iron is passivated by immersion in concentrated nitric acid; the initially vigorous attack of the metal ceases after a few seconds. Many other metals assume the passive state after comparable treatment.

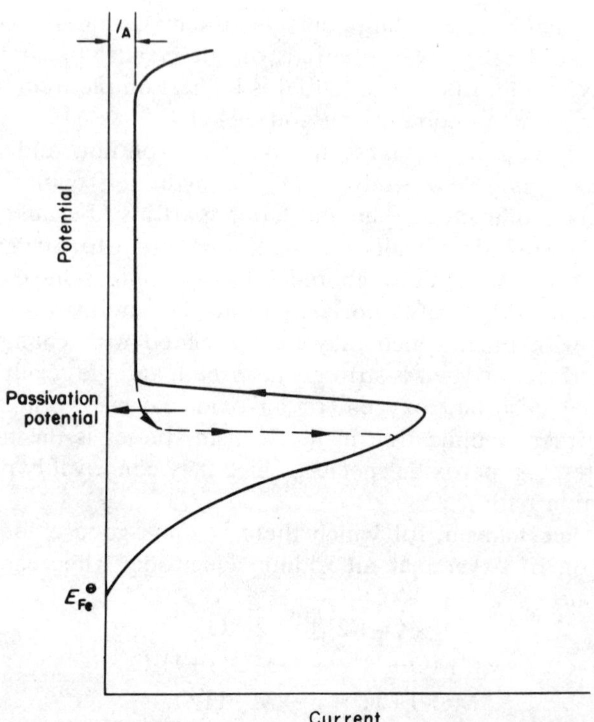


Figure P.1 Passivation curve ($I_A$ = anodic protection current)

Electrolytic passivation is illustrated in figure P.1, which shows the behaviour of an iron anode immersed in any suitable electrolyte solution when an increasingly positive potential is imposed on it. At first the current increases in the usual way as metal ions go into solution, but at the 'passivation potential' there is an abrupt change, and the current falls to a very low value which does not increase until some other anodic reaction takes over. The curve can be retraced, and the metal reactivated by holding it at a lower potential. Unless the potential change is very slow, the curve will show some hysteresis while the metal surface reverts to its original condition. Some passivation potentials, on the hydrogen scale, are: iron +0.58 V, nickel +0.36 V, chromium −0.22 V. Anodic protection, as a method of control of *corrosion* (q.v.), consists in passivating the metal surface and maintaining it thus by imposing an external e.m.f. which supplies the small current passing under these conditions (a current very much less than that required for 'cathodic protection').

Passivity has been attributed, since the time of Faraday, to an oxide film. In a few cases the presence of this has been directly demonstrated, and the response of a passive electrode to pH changes is very much that of a metal–metal oxide electrode. Uncertainties still exist, however, as to the nature and thickness of the surface film. On platinum, nickel and iron the onset of passivation may only require a monolayer of oxide, or of adsorbed oxygen atoms, although a layer several Ångstroms thick may be produced in time. Evidence for these very thin layers, which are quite undetectable visibly, comes from the very small cathodic pulse of electricity that is sufficient to remove them. In other cases, e.g. lead, quite thick films are needed to preserve the passive state. Partial protection is obtained for many metals in a variety of electrolytes so long as conditions favour the precipitation of a film of insoluble hydroxide or salt (see *Anodising*).

These differences depend mainly on the nature of the surface layer. If its conduction is electronic, an anodic reaction such as oxygen evolution can proceed freely at its surface, leaving the very thin oxide film unchanged. This is one extreme case; the other is that the film is porous to ions, in which case the attack of the metal will continue at a slow rate and the oxide film may grow to visible proportions.

*See also* E.

**Permittivity**

If two parallel conducting plates are electrified to surface charge densities $+\sigma$ and $-\sigma$, respectively, the field intensity between them, *in vacuo*, is given by

$$E_{vac} = 4\pi\sigma/\varepsilon_0 \qquad (P.1)$$

where $\varepsilon_0$ is the permittivity of a vacuum. When the space between the plates is occupied by some insulating material, the field strength is reduced, and is given by

$$E = 4\pi\sigma/\varepsilon \qquad (P.2)$$

where $\varepsilon$ is always greater than $\varepsilon_0$. The permittivity (units: $F\,m^{-1}$, $kg^{-1}\,m^{-3}\,s^4\,A^2$) of a material medium is related to the permittivity of a vacuum by

$$\varepsilon = \varepsilon_r \varepsilon_0$$

where $\varepsilon_r$ is the relative permittivity, also known as the dielectric constant of the medium—these are dimensionless quantities. ($\varepsilon_0 = 8.854 \times 10^{-12}\,F\,m^{-1}$.)

## Per-salts

Anodic oxidation in aqueous solution often provides the best method of obtaining an element in its highest oxidation state. Ammonium and potassium persulphates (peroxydisulphate) are prepared by electrolysing concentrated solutions of the corresponding sulphates with cathodes usually of lead and smooth platinum anodes (at which the oxygen overpotential is high). The over-all reaction is

$$2SO_4^{2-} \rightarrow S_2O_8^{2-} + 2e$$

but the system does not give a reversible potential, and several steps are probably involved.

Sodium perchlorate is made in a similar way by electrolysing concentrated sodium chlorate solutions between an iron cathode and a platinum anode:

$$ClO_3^- + H_2O \rightarrow ClO_4^- + 2H^+ + 2e$$

Potassium permanganate is made by electrolysing potassium manganate between steel cathodes and nickel anodes:

$$MnO_4^{2-} \rightarrow MnO_4^- + e$$

The cathode reaction gives $H_2$ and KOH, and the latter can be re-utilised in the preparation of more $MnO_4^{2-}$ from the starting material, manganese dioxide.

## pH

The pH of a solution is formally defined as the negative logarithm (to base 10) of the hydrogen activity of the solution:

$$pH = -\log a(H_3O^+) \tag{P.3}$$

Although the potential of hydrogen ion-indicating electrodes (hydrogen, glass, etc.) depends on $a(H_3O^+)$, the impossibility of measuring an individual electrode potential precludes the determination of the pH value of a solution as defined. The modern definition of pH is an operational one, in which a suitable indicator electrode is combined with a reference electrode, and the e.m.f. of the cell so formed is determined. The e.m.f. values, $E(X)$ and $E(S)$, of the cells

$$\ominus \ \mathrm{Pt}, \mathrm{H_2(g, 1\ atm)} \mid \text{Solution X} \vdots \text{KCl solution} \mid \mathrm{Hg_2Cl_2, Hg} \ \oplus$$

$$\ominus \ \mathrm{Pt}, \mathrm{H_2(g, 1\ atm)} \mid \text{Solution S} \vdots \text{KCl solution} \mid \mathrm{Hg_2Cl_2, Hg} \ \oplus$$

are measured at the same temperature; the molality of the bridge KCl solution should exceed 3.5 mol kg$^{-1}$. The e.m.f. values are

$$E(\mathrm{X}) = E(\mathrm{cal}) - E(\mathrm{H^+, H_2}) + E_1(\mathrm{X}) = E(\mathrm{cal}) - \frac{RT}{F} \ln a(\mathrm{H_3O^+})_x + E_1(\mathrm{X})$$

$$E(\mathrm{S}) = E(\mathrm{cal}) - E(\mathrm{H^+, H_2}) + E_1(\mathrm{S}) = E(\mathrm{cal}) - \frac{RT}{F} \ln a(\mathrm{H_3O^+})_s + E_1(\mathrm{S})$$

where $E$(cal) is the electrode potential of the reference *calomel electrode* (q.v.). Ignoring the two liquid junction potentials, $E_1(X)$ and $E_1(S)$, and subtracting these equations gives

$$E(\mathrm{X}) - E(\mathrm{S}) = \frac{RT}{F} \ln a(\mathrm{H_3O^+})_s - \frac{RT}{F} \ln a(\mathrm{H_3O^+})_x$$

$$= \frac{2.303\,RT}{F} [\mathrm{pH(X)} - \mathrm{pH(S)}]$$

or

$$\mathrm{pH(X)} = \mathrm{pH(S)} + \frac{F[E(\mathrm{X}) - E(\mathrm{S})]}{2.303\,RT} \qquad \text{(P.4)}$$

Now that the difference in pH of two solutions has been defined, the definition of pH(S) can be completed by assigning, at each temperature, a value of pH(S) to one or more chosen solutions designated as standards. The number assigned to pH(X) may or may not have any theoretical significance. It has significance when $E_1(X) = E_1(S)$; this is believed to be true when $I < 0.1$ mol kg$^{-1}$ and the pH is between 2 and 12. Under these conditions, the pH number has approximately the significance given by the formal definition.

Except in fairly concentrated acid solutions, hydrogen ion activities are nearly equal to the hydrogen ion concentrations. Hence, for normal working conditions,

$$pH = -\log c(H_3O^+)$$

since $a(H_3O^+) \times a(OH^-) = 10^{-14}$ mol$^2$ dm$^{-6}$ in any dilute solution, at neutrality $a(H_3O^+) = a(OH^-) = 10^{-7}$ mol dm$^{-3}$, and, hence, the pH value of a neutral solution is 7.0. Acid solutions have pH<7 and alkaline solutions have pH>7. For a 0.01 mol dm$^{-3}$ solution of a strong acid, pH = 2; for a weak acid the pH depends on the concentration and the acid dissociation constant:

$$pH = \tfrac{1}{2}pK_a - \tfrac{1}{2}\log\,(c/\text{mol dm}^{-3})$$

thus, the pH of a 0.1 mol dm$^{-3}$ solution of acetic acid, of $pK_a = 4.756$, is 2.88.

pH values in non-aqueous solution, obtained from electrometric determinations (as for aqueous solutions), must be regarded simply as numerical values which may be of practical use for their reproducibility. Equations for the e.m.f. value of cells containing aqueous solutions are not applicable to non-aqueous solutions for the following reasons: (a) high junction potential (e.g. at solvent–KCl boundary) between solvents of different physical and chemical properties becomes less stable and less reproducible as the concentration of water decreases; (b) solutions have high electrical resistance which makes potentiometric measurements less sensitive; and (c) the difficulty of choice of a suitable reference electrode. The only satisfactory method is to consider all solvents as independent systems, without making any reference to aqueous systems. The hydrogen electrode in any solvent is always accepted as the standard electrode, while the reference electrode (usually a calomel electrode; a silver–silver chloride electrode is of limited use) contains the same non-aqueous solvent which is being used. The value of the 'pH' obtained has a purely conventional significance in that $E^\ominus(H^+, H_2) = 0$ for all solvents, whereas in fact it almost certainly varies with the solvent. Glass, quinhydrone or antimony electrodes may be used in place of the hydrogen electrode.

*Measurement of pH*

(1) *Electrometric methods.* Table P.1 summarises the characteristics, limitations and reproducibility of various hydrogen ion-indicating electrodes.

(a) *Hydrogen electrode* (q.v.) and *calomel electrode* (q.v.):

$$\ominus\ \mathrm{Pt}, \mathrm{H_2(g, 101\,325\ N\ m^{-2})} \mid \text{Solution} \,\vdots\, \underset{\text{aq}}{\mathrm{KCl}} \mid \mathrm{Hg_2Cl_2, Hg}\ \oplus$$

$$E = E(\text{cal}) + 0.0591\ \text{pH}$$

(b) *Quinhydrone electrode* (q.v.) and calomel electrode:

$$\ominus\ \mathrm{Hg, Hg_2Cl_2} \mid \mathrm{KCl\ aq} \,\vdots\, \text{Solution, saturated with quinhydrone} \mid \mathrm{Pt}\ \oplus$$

$$E = E(\mathrm{Q, QH_2}) - E(\text{cal}) = E' - 0.0591\ \text{pH}$$

(c) *Antimony–antimony oxide electrode* (q.v.) and calomel electrode:

$$\ominus\ \mathrm{Sb, Sb_2O_3} \mid \text{Solution} \,\vdots\, \mathrm{KCl\ aq} \mid \mathrm{Hg_2Cl_2, Hg}\ \oplus$$

$$E = E' + 0.0591\ \text{pH}$$

Cell must be calibrated, $E'$ varies with ionic strength.

(d) *Glass electrode* (q.v.) and calomel electrode:

$$\ominus\ \mathrm{Ag, AgCl} \mid \mathrm{HCl\ (0.1\ mol\ dm^{-3})} \mid \text{glass} \mid \text{solution} \,\vdots\, \mathrm{KCl\ aq} \mid \mathrm{Hg_2Cl_2, Hg}\ \oplus$$

←——— glass electrode ———→

Since the glass has such a high resistance, the e.m.f. of this cell cannot be measured directly with a *potentiometer* (q.v.); instead a *pH meter* (q.v.) with a valve or transistor high-output impedance circuit must be used. Because of the asymmetry potential of the glass electrode, which varies with time, the pH meter must be standardised regularly with solutions of known pH.

(2) *Indicator methods.* For an acid–base *indicator* (q.v.),

$$\underset{\substack{\text{acidic form}\\ \text{colour A}}}{\mathrm{HIn_A}} + \mathrm{H_2O} \rightleftharpoons \underset{\substack{\text{basic form}\\ \text{colour B}}}{\mathrm{In_B^-}} + \mathrm{H_3O^+}$$

$$K_{\mathrm{In}} = \frac{c(\mathrm{In_B^-})\,c(\mathrm{H_3O^+})}{c(\mathrm{HIn_A})}$$

Table P.1 Characteristics of electrodes used for measurement of pH

| Electrode | pH range | Interfering substances | Causes of error | Reproducibility/mV | Remarks |
|---|---|---|---|---|---|
| Hydrogen | 0–14 | Oxidising and reducing agents, air, heavy metals | Incomplete saturation, $O_2$ in $H_2$, poisons | 0.1 | Follows theoretical equations, slow attainment of equilibrium, strong reducing action |
| Quinhydrone | 1–8 | Alkali, oxidising and reducing agents, complexing agents, proteins | Poisons, salts | 0.1 | Follows theoretical equation, rapid attainment of equilibrium, solution contaminated |
| Antimony–antimony oxide | 3–10 | Strongly acid or alkaline conditions, $H_2S$, $Cu^{2+}$ | Oxidising agents organic compounds, salts, $Cl_2$, $H_2S$ | 5 | Does not follow theoretical relationship, must be calibrated |
| Glass | 0–12 | Dehydrating agents, colloids, surface deposits | High concentration of alkali | 0.5 | Follows theoretical equation over certain pH range depending on type of glass, must be calibrated, equilibrium readily established, can be used in the presence of oxidising and reducing agents |

whence,
$$\mathrm{pH} = \mathrm{p}K_{\mathrm{In}} + \log \frac{c(\mathrm{In}_{\mathrm{B}}^{-})}{c(\mathrm{HIn}_{\mathrm{A}})}$$

$$= \mathrm{p}K_{\mathrm{In}} + \log \frac{\text{intensity of colour B}}{\text{intensity of colour A}}$$

assuming that the concentrations of the two forms are proportional to the intensities of light transmitted by solutions of colours A and B. Thus, provided that $\mathrm{p}K_{\mathrm{In}}$ is known, the pH of a solution can be obtained from a measurement of the ratio of intensities of the colours using a visual comparator or a photoelectric spectrophotometer.

In an alternative indicator method the colour of the indicator in the solution of unknown pH is compared with that of standard buffer solutions containing the same concentration of indicator.

*See also* J & P, Mi, R & S, V.

**pH meter**

The potential of a cell which includes a *glass electrode* (q.v.) cannot be measured on a simple potentiometer, since the glass electrode has a resistance of $10^{7}$–$10^{8}$ $\Omega$. For an accuracy of 0.2 pH units, the electrode potential must be measured to better than 1 mV; this means that the instrument used for detecting the balance point of the potentiometer must be sensitive to at least $10^{-10}$ A. For such measurements the electrodes must be connected to a special electrometer valve which has a very high input resistance. This was the basis of the original null-point pH meters.

The modern mains-operated direct-reading pH meters are fundamentally electronic millivoltmeters with an exceptionally high input resistance. The basic stability of such instruments is such that day-to-day accuracy is within ±0.02 pH for a period of 24 hours. Continuous measurements can be made without frequent standardisation in buffer solutions; the calibration can be readily checked without disturbing the electrodes (see table A.V).

A change in the temperature of a solution affects the electrode system in two ways; firstly, the pH/e.m.f. relationship alters, and, secondly, there is a zero shift in the pH scale. Such instruments usually contain a fully automatic temperature device which ensures correct readings at all temperatures. In addition to the glass and calomel reference electrodes, a resistance thermometer is also immersed in the solution.

**Pitt's conductance equation**
*See* Conductance equations.

**Platinum and gold electrodes**
Precious metal electrodes are used either as reference electrodes or for making electrical contact in oxidation–reduction systems. Platinum, the more commonly used metal, may be used in the shiny form (oxidation–reduction electrodes), the black form (hydrogen electrode and conductance cells to reduce polarisation errors) or the grey form (conductance cells).

Platinum black or platinised platinum electrodes, with an increased surface area, are obtained by depositing a layer of platinum on the surface of a platinum electrode. A suitable platinising solution contains 3 g of chloroplatinic acid and 0.025 g of lead acetate dissolved in 100 $cm^3$ of water. The electrodes, after stringent cleaning, are immersed in this and a current sufficient to cause a moderate rate of gassing is passed for 10–15 min; its direction is reversed every 30 s. A moderately thick coat of platinum black is preferable to a thin one since it gives a greater surface area and is not so readily eroded away. The deposit contains occluded gas and liquid. These are removed by immersing the electrodes in sulphuric acid (0.3 mol $dm^{-3}$) and connecting them through a reversing switch to a 4 V accumulator. The solution is electrolysed for 30 min with reversal of polarity every 30 s, so that gas bubbles freely from both electrodes. The black coating may be stripped off by electrolysing either aqua regia or 5 mol $dm^{-3}$ hydrochloric acid, with the electrode connected to the positive pole of an accumulator.

In exceptional cases, such as with certain coordination compounds, platinum black has objectionable catalytic effects and bright or 'greyed' electrodes should be used. Grey electrodes are prepared by heating black electrodes to a dull red heat; this reduces the surface area. In very dilute solutions blacked electrodes may cause a drift in measured conductance readings because of adsorption of the electrolyte from solution.

*See also* Conductance, electric.

**Polarisable electrodes**

An ideal non-polarisable electrode is a system through which large currents can pass in either direction without displacing the electrode potential from its equilibrium position. This is unattainable in practice, but electrode systems which have a high *exchange current density* (q.v.), such as the $Ag^+$, Ag electrode, can support large currents at quite a small *overpotential* (q.v.).

The opposite extreme, the ideal 'polarisable electrode', is one through which no current would pass whatever potential were imposed on the electrode. This condition is approached within certain potential limits by electrode–solution combinations in which no reversible charge-transfer process is established.

**Polarisation**

An electrode is said to be polarised whenever its potential departs from its equilibrium value.

*See also* Concentration overpotential; Polarisable electrodes.

**Polarography**

The technique of polarography has developed in several directions since it was first introduced by Heyrovsky; the essentials of the method are illustrated in figure P.2. A is the cell containing the solution, and dipping into this is a fine capillary tube, at the tip of which droplets of mercury form at a controlled slow rate and fall through the solution. The growing drops of mercury form the electrode, usually the cathode, at which the reaction to be studied takes place. The other electrode is the pool of mercury at the bottom of the vessel, which can be assumed to be at a constant potential throughout the measurement; a standard calomel electrode can be used in its place as the second electrode. The solution is swept out with nitrogen to remove the dissolved oxygen, which would otherwise be reduced at the cathode and interfere with the determination. During an experiment the voltage across the cell, and therefore the potential at the dropping mercury electrode, is increased continuously by means of the rheostat, R, while the current flowing is read from the galvanometer, G. The whole process can be made automatic to give a pen-recording of the current–potential diagram.

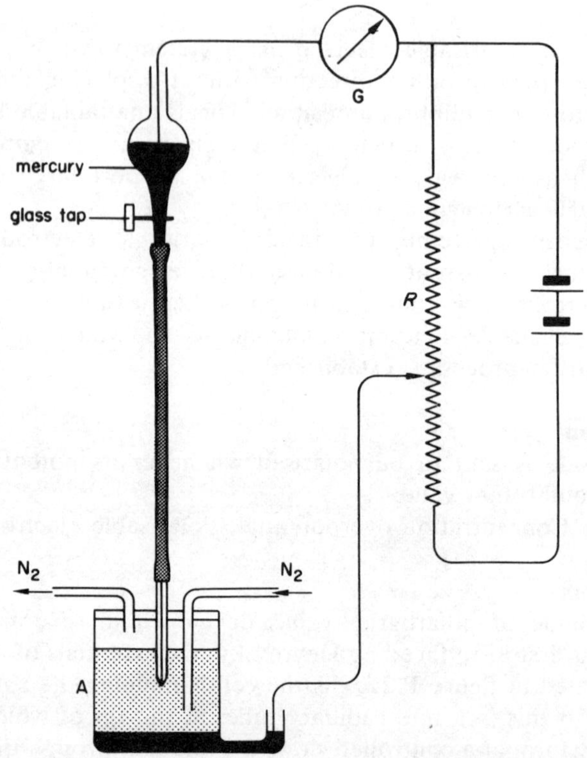


Figure P.2 Simple polarograph

Suppose a mixture of zinc and cadmium ions is to be analysed. The solution is arranged to contain the two ions at concentrations of the order of $10^{-3}$ mol dm$^{-3}$, and an *indifferent electrolyte* (q.v.), i.e. one that takes no part in the electrode reaction, such as $KNO_3$, at a concentration of about 1 mol dm$^{-3}$. This ensures that the solution has a high conductivity and that the indifferent electrolyte carries virtually the whole of the current so that the electroactive ions must arrive at the electrode surface by diffusion. Under these conditions, the polarogram should appear as in figure P.3. ABG is the residual current, which can be plotted directly from a parallel experiment with only

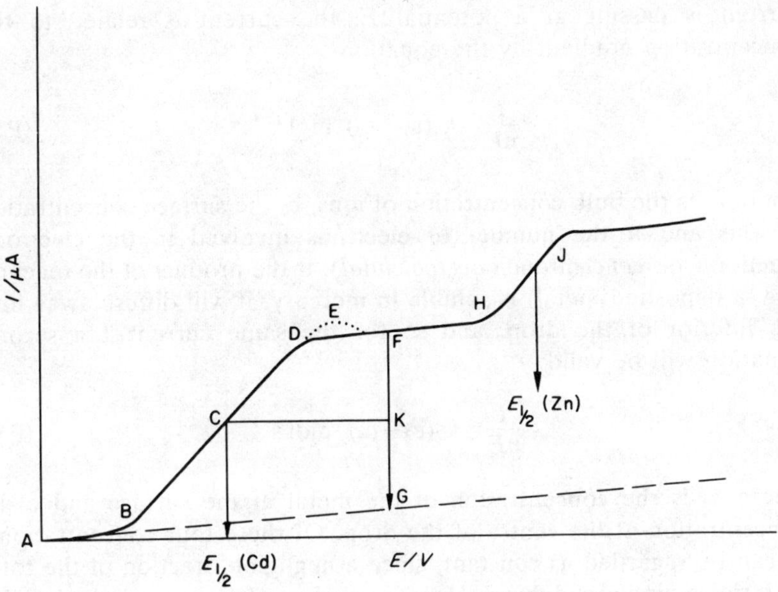


Figure P.3 Polarogram (see text)

1 mol dm$^{-3}$ $KNO_3$ in the cell. B is the decomposition potential of cadmium, and from B onwards cadmium is discharged at the dropping electrode. As the potential is increased, the limiting current for cadmium deposition is reached at D, and the curve becomes nearly horizontal, or parallel to the residual current line. At H the decomposition potential of zinc is reached, and a second polarographic wave is traced out while zinc and cadmium are deposited simultaneously. A spurious maximum (the dotted line through E) is commonly observed before the current returns to the theoretical line through F and H. This is attributed to adsorption phenomena, and is avoided by adding a small quantity of a 'maximum suppressor' such as gelatin or other surface-active material to the original solution.

The theoretical explanation of a polarographic wave involves four concentrations. The ion being discharged arrives at the mercury surface by diffusion, and considering steady state conditions in which a current is passing at a potential $E$, the current is related to the concentration gradient by the equation

$$\frac{I}{nF} = k_1(c_1 - c_2) \ \text{mol s}^{-1} \tag{P.5}$$

where $c_1$ is the bulk concentration of ions, $c_2$ the surface concentration of ions and $n$ the number of electrons involved in the electrode reaction (see *concentration overpotential*). If the product of the reaction (e.g. a deposited metal) is soluble in mercury, it will diffuse away into the interior of the drop, and so for the same current $I$ a second equation will be valid:

$$\frac{I}{nF} = k_2(c_3 - c_4) \ \text{mol s}^{-1} \tag{P.6}$$

where $c_3$ is the concentration of the metal at the surface and $c_4$ its concentration at the centre of the drop. Of these four concentrations, $c_1$ can be regarded as constant, since a negligible fraction of the total material is discharged during the course of a polarogram; and $c_4$ can be put equal to zero, as the mercury is being continuously removed. The two remaining concentrations will be connected by the Nernst equation for the potential:

$$E = E^{\ominus} + \frac{RT}{\text{nF}} \ln\,(\gamma_2 c_2/\gamma_3 c_3) \tag{P.7}$$

where $\gamma_2$ and $\gamma_3$ are the activity coefficients in the solution and in the amalgam, respectively. *Activation overpotential* (q.v.) is here assumed to be negligible.

As the current is increased, $c_2$ must decrease (equation P.5) and must eventually approach zero; the current passing is now the maximum current that the process can support, and $E$ rises abruptly to a value at which some other electrode process is possible. This is the limiting current (see *limiting current density*) and equation (P.5) becomes

$$\frac{I_{\text{lim}}}{nF} = k_1 c_1 \ \text{mol s}^{-1} \tag{P.8}$$

From equation (P.6), with $c_4 = 0$,

$$c_3 = \frac{I}{nF}\frac{1}{k_2} \tag{P.9}$$

and from equations (P.5) and (P.8),

$$\frac{I}{nF} = \frac{I_{\text{lim}}}{nF} - k_1 c_2$$

or

$$c_2 = (I_{\text{lim}} - I)/nFk_1 \tag{P.10}$$

When these values are substituted in equation (P.7),

$$\begin{aligned} E &= E^{\ominus} + \frac{RT}{nF}\ln\frac{\gamma_2(I_{\text{lim}} - I)k_2}{\gamma_3 I k_1} \\ &= E^{\ominus} + \frac{RT}{nF}\ln\frac{\gamma_2 k_2}{\gamma_3 k_1} + \frac{RT}{nF}\ln\frac{(I_{\text{lim}} - I)}{I} \end{aligned} \tag{P.11}$$

When the current is half the limiting current, equation (P.11) becomes

$$E_{\frac{1}{2}} = E^{\ominus} + \frac{RT}{nF}\ln\frac{\gamma_2 k_2}{\gamma_3 k_1} \tag{P.12}$$

$E_{\frac{1}{2}}$ is found by bisecting the vertical line drawn from the limiting current to the residual current line, FG in figure P.3. FG is the wave height, and a horizontal line from K gives C and thus the half-wave potential, $E_{\frac{1}{2}}$. $E_{\frac{1}{2}}$ has a characteristic value for an electrode reaction, and can be used for qualitative analysis. It is related to the standard electrode potential, $E^{\ominus}$, by equation (P.12), which involves the activity coefficients and the diffusion characteristics in the solution and in the growing drop. Activation overpotential may also contribute to $E_{\frac{1}{2}}$, and it is best regarded as an independent constant for the reaction, which can be determined in separate calibration experiments.

The wave-height is the quantity required for quantitative analysis, since by equation (P.8) it is proportional to the bulk concentration of ions. Again it is more accurate, however, to estimate the concentration by comparison with calibrating runs made with known quantities of the substance.

A complication not yet mentioned is that the surface area of the electrode fluctuates as one drop of mercury expands and is replaced by another. It has been shown, however, that the theory outlined is valid for the mean current. This is given, according to the Ilkovic equation, by

$$I_{\text{lim}} = 60.70\, w^{2/3} t^{1/6} D^{1/2} z_i c_0$$

where $w$ is the rate of flow of mercury in g s$^{-1}$, $t$ is the life-time of the drop, and $D$, $z_i$ and $c_0$ the diffusion coefficient, charge number and concentration of the reacting ion, respectively.

A rotating platinum disc may be used instead of the dropping mercury electrode. With rapid rotation, high and stable limiting currents can be obtained, and oscillographic records obtained with this electrode can provide information about very rapid reactions.

*See also* K & L; and Zuman, P. (1964), *Organic Polarographic Analysis*, Pergamon.

**Potential, reversible electrode**
*See* Electrode potential.

**Potentiometer**
The e.m.f. of any *cell* (q.v.) may be measured using a simple potentiometer, a valve voltmeter (e.g. pH meter) or a *digital voltmeter* (q.v.). A standard *Weston cell* (q.v.) of known e.m.f. is used for the calibration of a potentiometer.

A steady source of e.m.f. (accumulator) is used to establish a potential gradient along AB (figure P.4). The potential drop between

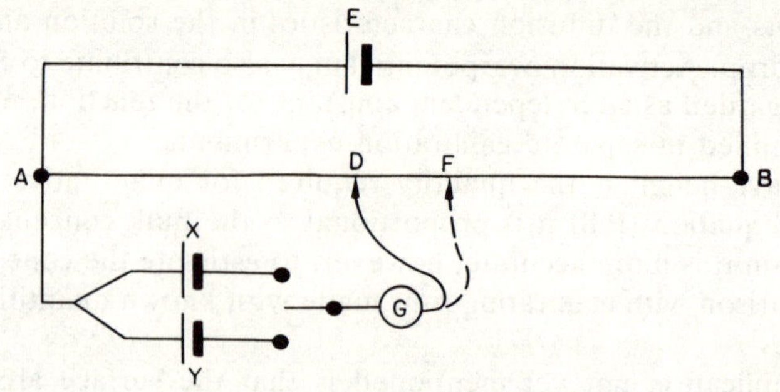


Figure P.4 Simple wire potentiometer

A and D depends on the resistance of the wire between these points. If a cell, X, of e.m.f. $E(X)$ is connected through the galvanometer, G, to the sliding contact D on AB there will be no flow of current if the p.d. from A to D is equal to $E(X)$. Thus the position of D can be found experimentally. In a similar way for cell, Y, when the p.d. between A and F is equal to $E(Y)$, the galvanometer deflection is zero. Hence,

$$\frac{E(X)}{E(Y)}=\frac{\text{p.d. from A to D}}{\text{p.d. from A to F}}=\frac{\text{resistance of AD}}{\text{resistance of AF}}=\frac{\text{AD}}{\text{AF}}$$

The e.m.f. of a cell may be determined from a knowledge of the ratio of the resistances AD/AF, for the cell and a standard cell of known e.m.f. In its simplest form, with a wire of uniform dimensions and resistance, the ratio of the resistances is equal to the ratio of the lengths.

In commercial potentiometers the metre wire of the simple potentiometer is replaced by a series of coils of standard resistance and a graduated slide wire. The potential across these resistances is adjusted by suitable calibration of the accumulator against a standard Weston cell. Once standardised, the unknown e.m.f. is balanced with the selector switch in the 'read' position. The cell must not be kept permanently in circuit; the tapping key is depressed momentarily. Some instruments are provided with a range switch which permits greater accuracy in the measurement of small e.m.f. values.

## Potentiometric titration

In a potentiometric titration the variation of the potential of an electrode in equilibrium with its ions indicates the course of the reaction. Precise location of the end-point is obtained by extrapolation from a series of independent observations rather than from one estimate of the end-point as with indicators.

The *electrode potential* (q.v.) between a metal and a solution of one of its salts, $M \rightleftharpoons M^{n+} + ne$, is (neglecting activity coefficients)

$$E(M^+, M) = E^{\ominus\prime}(M^+, M) + \frac{RT}{nF} \ln c(M^+)$$

and for a redox system

$$E(O, R) = E^{\ominus\prime}(O, R) - \frac{RT}{nF} \ln \frac{c(\text{ox})}{c(\text{red})}$$

where $E^{\ominus\prime}(O, R)$ is the formal standard redox potential (i.e. the electrode potential when $c(ox) = c(red)$), which is effectively constant for a given ionic strength such as obtains during a titration. The variation of $E(M^+, M)$ or $E(O, R)$ thus reflects changes in $c(M^+)$ and $c(ox)/c(red)$, respectively. Since it is impossible to measure a single electrode potential, the indicating electrode must be used in conjunction with a reference electrode, the potential of which does not change during the titration. Often, owing to chemical reaction, it may be necessary to interpose a *salt bridge* (q.v.) between the reference electrode and the solution being titrated. Thus, for an acid–base titration, a suitable cell would be

$$\ominus\ \mathrm{Pt, H_2} \left| \begin{array}{c}\text{Solution}\\ \text{to be}\\ \text{titrated}\end{array} \right\vdots\ \mathrm{KCl} \left| \mathrm{Hg_2Cl_2, Hg}\ \oplus \right.$$

for which, at 298 K,

$$E(\text{cell}) = E(\text{cal}) - E(H^+, H_2) = E' + 0.0591\ \text{pH}$$

In a typical potentiometric titration, the initial addition of a small amount of titrant produces little change in the e.m.f. of the cell (figure P.5), since this depends on the fraction of a particular ion removed. Towards the equivalence point, however, the fraction of ion removed by a constant amount increases rapidly, and this is reflected by a rapid change in e.m.f. Above the equivalence point the curve again flattens out. Graphical procedures for the exact location of the equivalence point are (a) method with first derivative (figure P.6a), (b) method with second derivative (figure P.6b) and (c) Gran's method, $\dfrac{\Delta V}{\Delta E/V^{1/2}}$ against $V$ (figure P.6c).

The following conditions must be fulfilled by an analytical reaction which is to be used potentiometrically:

(1) the substance to be titrated and the titrant must react in a fixed known stoichiometric ratio;

(2) an indicator electrode must be available, the electrode potential of which must be uniquely and reversibly determined by the concentration of one ion in the reaction;

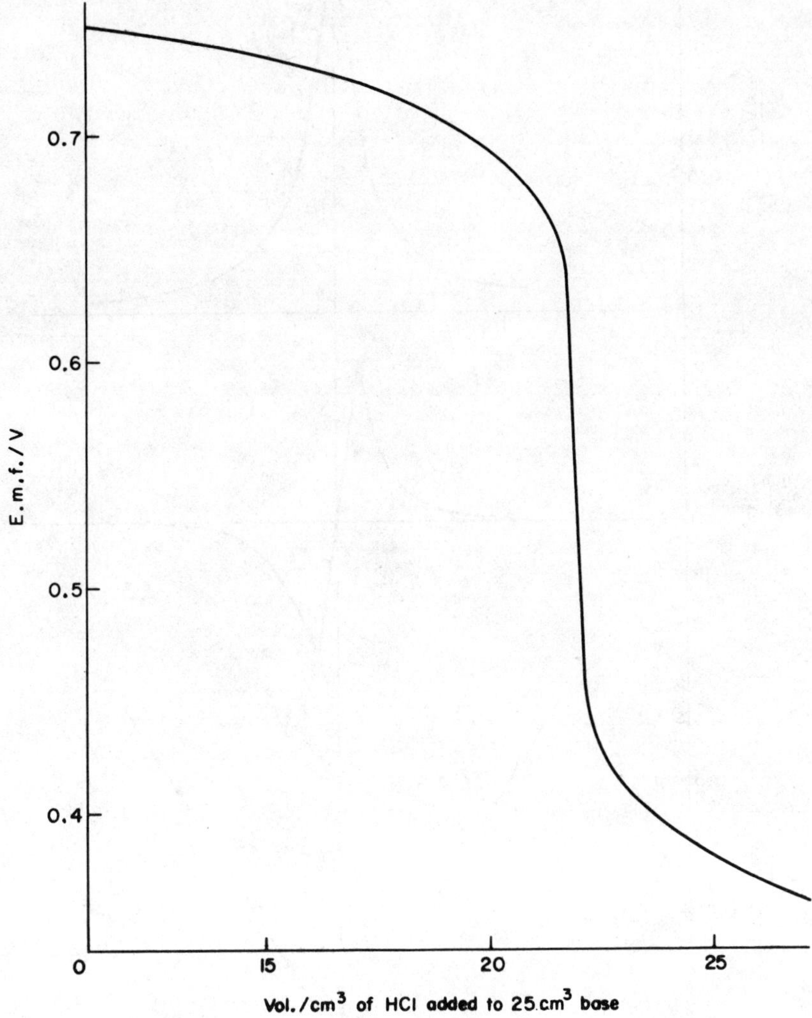


Figure P.5 Typical titration curve of weak base against strong acid (using hydrogen and saturated calomel electrodes)

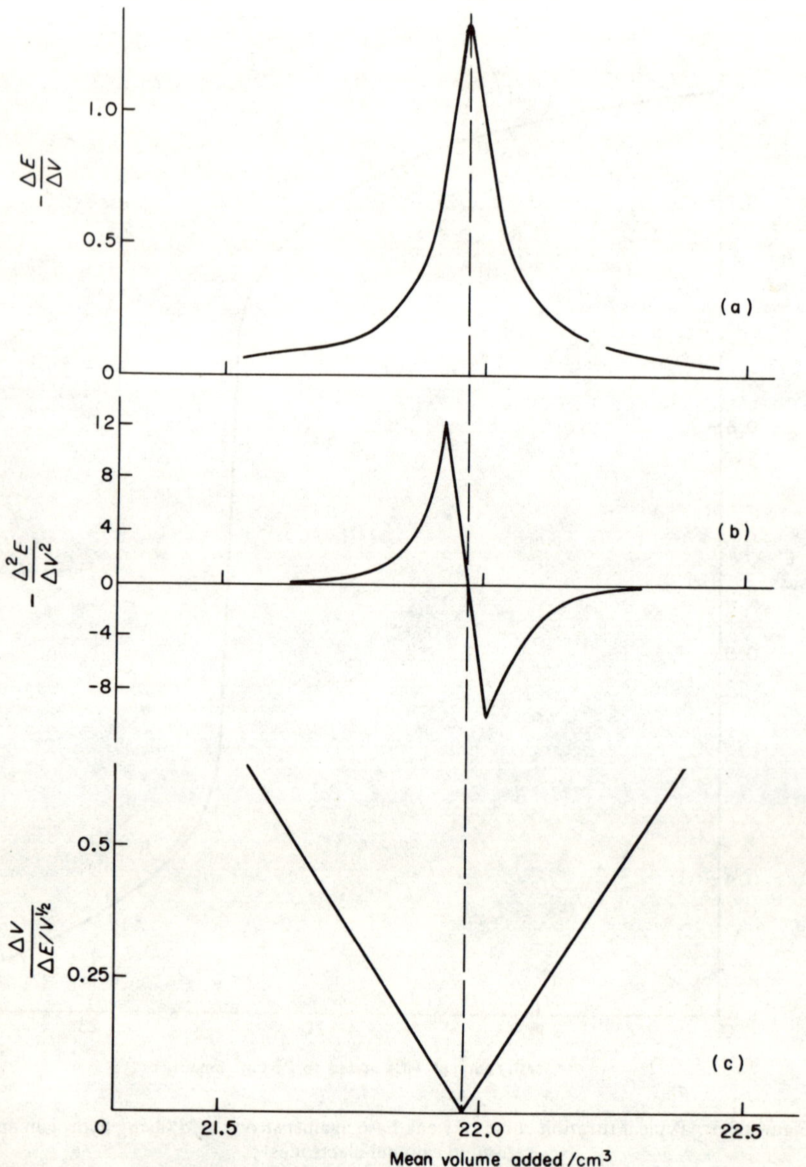


Figure P.6 Differential titration curve. (a) First differential titration curve. (b) Second differential titration curve. (c) Gran's plot

(3) the reaction must be as complete as possible, i.e. as high an equilibrium constant as possible;
(4) rapid equilibrium must be established between reacting components and the indicator electrode.

*Acid–base titrations*

The shape and position of the titration curve (figure P.7) depends on the strengths of acid and base. The curve for the titration of a strong acid with a strong base (I–II) is readily calculated since $c(H^+)$ = concentration of unneutralised acid up to the equivalence point (B). At this point the solution is identical with one containing the neutral salt, and $c(H^+) = 10^{-7}$ mol dm$^{-3}$. Further addition of base results in the solution containing free base, and the pH can be readily calculated.

The curve for the titration of a weak acid with a strong base (II–III) can be calculated up to the equivalence point (C) with reasonably good results by the *Henderson equation*‡ (q.v.):

$$\mathrm{pH} = \mathrm{p}K_a + \log \frac{c(\mathrm{A}^-)}{c(\mathrm{HA})}$$

At the end-point the solution is alkaline owing to hydrolysis of the salt

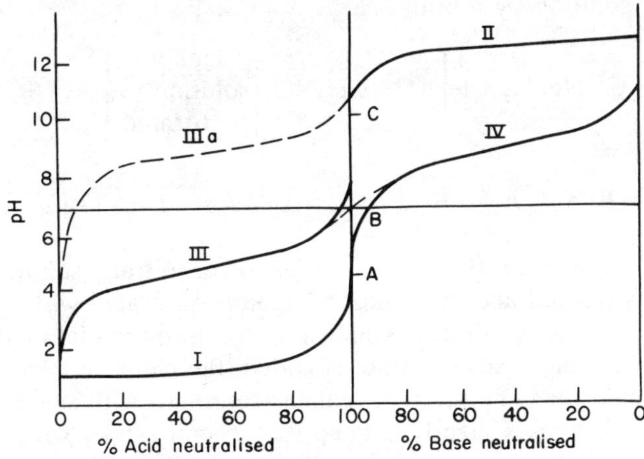


Figure P.7 Neutralisation curves for acids and bases (all concentrations 0.1 mol dm$^{-3}$): I, strong acid; II, strong base; III, weak acid; IIIa, very weak acid; IV, weak base. A is equivalence point for I–IV; B for I–II and C for II–III

of a weak acid and strong base; the pH is given by

$$pH = \tfrac{1}{2}pK_w + \tfrac{1}{2}pK_a + \tfrac{1}{2}\log c$$

Similarly, the curve for the titration of a weak base with a strong acid (I–IV) can be calculated; the pH at the equivalence point, A, is given by

$$pH = \tfrac{1}{2}pK_w - \tfrac{1}{2}pK_b - \tfrac{1}{2}\log c$$

The exact treatment of the titration of a weak acid with a weak base (III–IV) is more complicated; it is analogous to that for the hydrolysis of the salt of a weak acid and weak base.

Comparing the four titration curves, there is a sharp change of pH at the equivalence point for a strong acid–strong base, a smaller but usually measurable change for a weak acid–strong base or strong acid–weak base; but for a weak acid–weak base the change of pH is very gradual throughout the titration, and is not at all marked at the equivalence point.

*Precipitation titrations*

These are generally limited to those using silver nitrate as one reagent with a silver indicating electrode, and a reference electrode separated from the solution by a bridge, e.g.

$$\ominus\ Hg, Hg_2Cl_2 \,\Big|\, KCl \,\vdots\, NH_4NO_3 \,\vdots\, \begin{matrix}\text{Halide}\\ \text{solution to}\\ \text{be titrated}\end{matrix} \,\Big|\, Ag\ \oplus$$

for which

$$E(\text{cell}) = E(Ag^+, Ag) - E(\text{cal}) = E^\ominus(Ag^+, Ag) - E(\text{cal}) + \frac{RT}{F}\ln a(Ag^+)$$

On the addition of the first drop of silver nitrate solution, silver chloride is formed and the solution becomes saturated with respect to it; $c(Ag^+)$ is thus small and is indicated by the potential of the silver electrode. As more silver nitrate is added, the chloride is precipitated as silver chloride. The solution remains saturated with silver chloride, but $c(Ag^+)$ increases slightly to keep $K_s$ constant as $Cl^-$ is removed. At the end-point $c(Ag^+)$ increases rapidly owing to the presence of excess silver ions, and the silver electrode shows a rapid increase of potential (figure P.8).

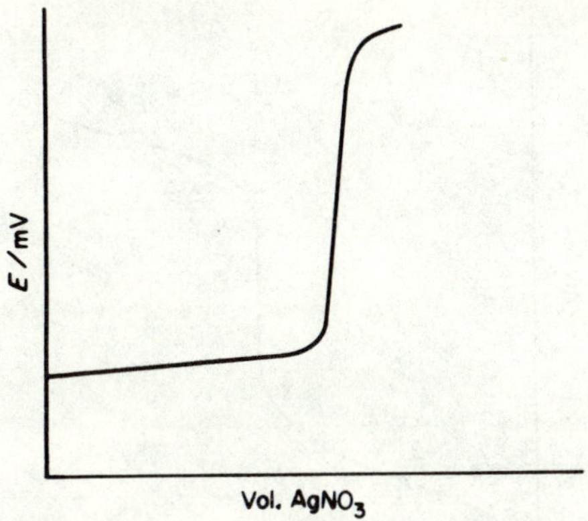


Figure P.8 Precipitation titration curve

With such a cell it is possible in a single titration to determine the concentrations of iodide, bromide, and chloride ions in solution, since the different solubility products are reflected in three distinct steps in the electrode potential (or e.m.f. of the cell), corresponding to the complete precipitation of AgI, AgBr and AgCl (figure P.9).

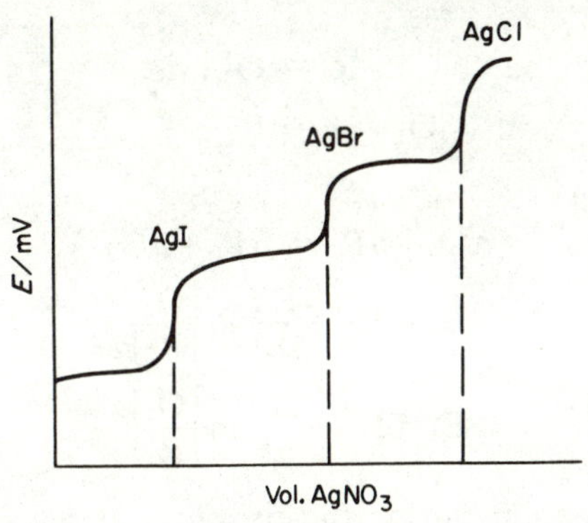


Figure P.9 Multiple precipitation potentiometric titration

*Redox titrations*

The position of the first part of the curve (figure P.10) is determined by the standard electrode potential of the titrated system and the second

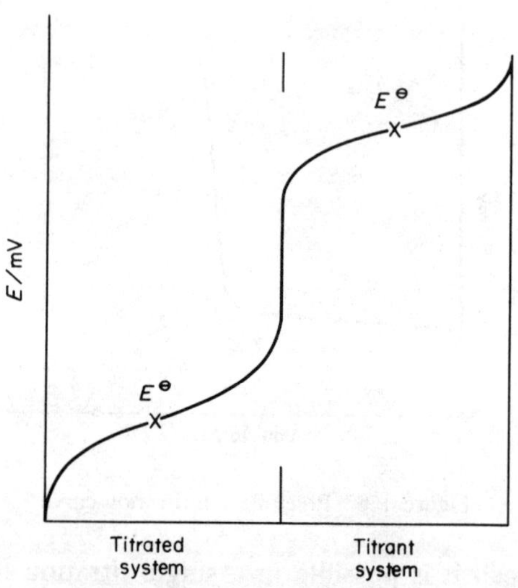


Figure P.10 Oxidation–reduction titration curve

half of the curve by the standard electrode potential of the titrant. For the general case of two redox systems involving different numbers of electrons:

$$R_2 \rightarrow O_2 + ae$$

$$O_1 + be \rightarrow R_1$$

the cell reaction is

$$aO_1 + bR_2 \rightleftharpoons aR_1 + bO_2$$

and

$$E_2 = E_2^{\ominus} + \frac{RT}{aF} \ln \frac{[O_2]}{[R_2]}$$

$$E_1 = E_1^{\ominus} + \frac{RT}{bF} \ln \frac{[O_1]}{[R_1]}$$

$[O_1] = [R_2]$ at any point during the titration; at the equivalence point $[O_2]$ is also equal to $[R_1]$, hence $[O_1]/[R_1] = [R_2]/[O_2]$. Thus at the equivalence point

$$E_e = E_1 = E_2 = E_2^{\ominus} + \frac{RT}{aF} \ln \frac{[R_1]}{[O_1]} = E_1^{\ominus} + \frac{RT}{bF} \ln \frac{[O_1]}{[R_1]}$$

from which it is apparent that

$$E_e = \frac{aE_2^\ominus - bE_1^\ominus}{a+b}$$

Further, since

$$K = \frac{[R_1]^a[O_2]^b}{[O_1]^a[R_2]^b}$$

it follows that

$$\frac{[R_1]}{[O_1]} = \frac{[O_2]}{[R_2]} = K^{1/(a+b)}$$

This is an expression that permits the calculation of the exact concentrations at the equivalence point and, hence, the feasibility of the titration.

In the titration of an iron(II) salt with potassium permanganate

$$5Fe^{2+} + MnO_4^- + 8H^+ \rightarrow 5Fe^{3+} + Mn^{2+} + 4H_2O$$

$$E_e = \frac{E^\ominus(Fe^{3+}, Fe^{2+}) - 5E^\ominus(MnO_4^-, Mn^{2+})}{6} = 1.39 \text{ V}$$

and $$\log K = \log \frac{[Mn^{2+}][Fe^{3+}]^5}{[MnO_4^-][Fe^{2+}]^5[H^+]^8} = \frac{(1.52-0.78)5}{0.059} = 63.5$$

At the equivalence point $[Fe^{3+}]/[Fe^{2+}] = (3\times 10^{63})^{1/6} = 4\times 10^{10}$. This large value of $K$ is brought about by the large difference in $E^\ominus(O, R)$ for the two systems. If a given reaction is to be suitable for volumetric analysis, it is necessary that $K > 10^6$. This corresponds to a minimum difference of $E^\ominus(O, R)$ of 0.35 V if $n = 1$ for both systems, 0.26 V if $n = 1$ and $n = 2$ for the two systems, and 0.18 V if $n = 2$ for both systems. Typical titrations are shown in figure P.10.

*Experimental methods*

(1) Direct titration using suitable indicator and reference electrodes. The e.m.f., or pH, is determined after addition and thorough mixing of measured aliquots of titrant. It is advantageous to make large additions of titrant (2 cm$^3$ to 25 cm$^3$ of titrated system) in the early stages, gradually decreasing the amount (0.1–0.05 cm$^3$) towards the end-point. The full titration curve and derivative curves can be plotted from these data.

*Potentiometric titration*

For acid–base titrations the glass and calomel electrode assembly is used with a pH meter, while for redox titrations cells of the type

$$\ominus \; Hg, Hg_2Cl_2 \left| \begin{matrix} KCl \\ sat \end{matrix} \right\vdots \; Fe^{2+}, Fe^{3+} \; \Big| \; Pt \; \oplus$$

are used. Depending on the type of system, it may be necessary to make the titration in an atmosphere of nitrogen.

(2) Differential method (applicable to acid–base, precipitation and redox titrations). In this method the differential curve is plotted directly instead of being calculated from the e.m.f.–volume graph. For the titration of solution AB with solution CD, X and Y are electrodes reversible to $A^+$, connected to the measuring instrument; Y is enclosed in a tube which temporarily holds back a portion of AB (figure P.11).

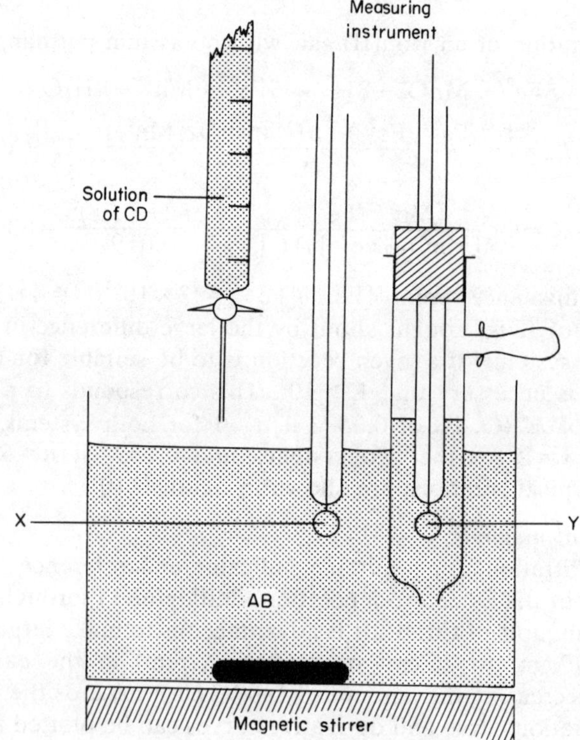


Figure P.11 Apparatus for differential titration

Initially $E(\text{cell}) = 0$, but on addition of CD the potential of X is changed, owing to the removal of $A^+$ from solution (e.g. by precipitation or neutralisation), while that of Y is unchanged since the trapped solution has not mixed with the bulk. The e.m.f. is recorded; this is $\Delta E$ for a given volume, $\Delta V$, of CD added. The sheltered solution is mixed with the bulk so that $E(\text{cell}) = \Delta E = 0$. This procedure is repeated for further additions of CD, and $\Delta E/\Delta V$ plotted against volume of titrant added; the maximum in the curve corresponds to the end-point.

(3) Bottled end-point method (applicable to acid–base, precipitation and redox titrations). In this method the reference electrode is replaced by a compensation electrode (figure P.12), the potential of which is exactly equal to that of the indicator electrode in the solution under titration at the end-point (previously determined). Electrodes reversible to one of the ions in the titration are coupled through a galvanometer so that the end-point is indicated by a sudden reversal of

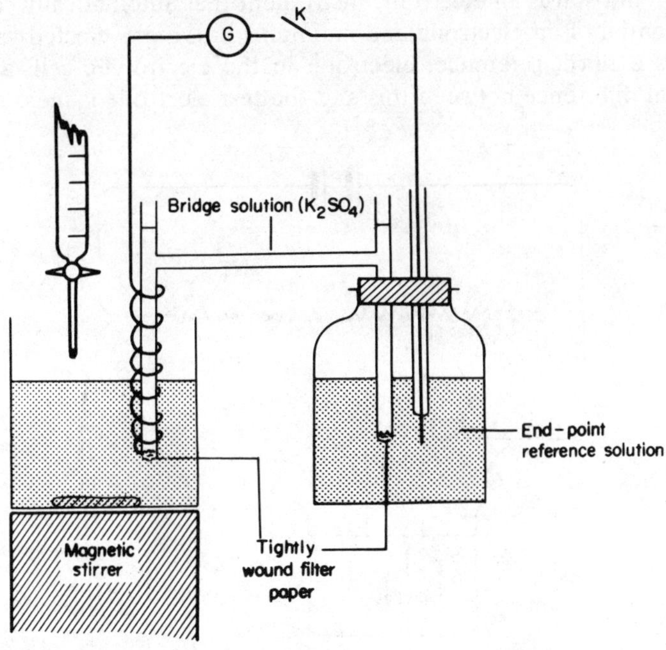


Figure P.12 Apparatus for bottled end-point titration

polarity. At the start of titration large galvanometer deflections are recorded; these gradually decrease to zero at the end-point, thereafter increasing in the opposite direction. This is a very simple method for determining the end-point, requiring only a sensitive galvanometer, and is ideal for routine work.

*See also* Amperometric titration; Conductimetric titration; Thermometric titration‡; and G, I & J, J & P, Mi, S, G & W, V.

**Potentiostat**

A simple circuit for electrolysis at an approximately constant voltage is illustrated in figure P.13. R is a rheostat of fairly low resistance, so that the current passing through the cell is negligible compared with the current flowing through the rheostat. Under these conditions the voltage applied to the electrodes A and C will be independent of small changes in the electrolysis current.

A potentiostat is an electronic instrument that automatically controls the potential of an electrode and maintains it at a preselected value. It requires a third, reference, electrode in the electrolytic cell, and the potential difference between this and the test electrode is measured by

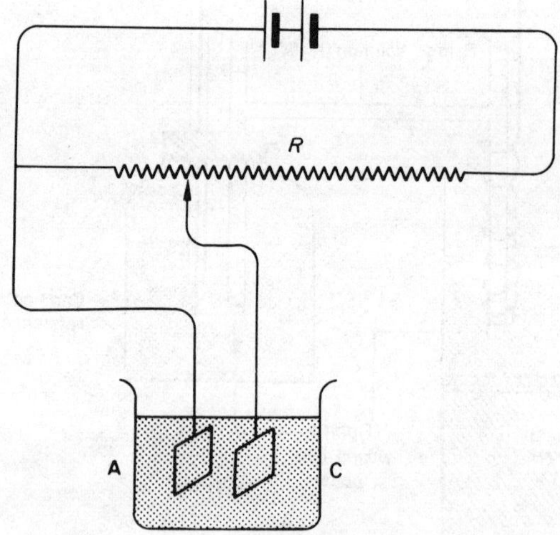


Figure P.13 Circuit for constant-voltage electrolysis

an electronic voltmeter. Whenever there is a discrepancy between the value so measured and the required value, the potentiostat is actuated to pass a current through the cell of the direction and magnitude required to remove the discrepancy.

**Primary cell**

A great variety of chemical reactions can be used as the basis for primary cells. Most practical arrangements will give voltages in the range 1–2 V, corresponding to reactions with a free energy change of about 210 kJ $mol^{-1}$. This, for instance, is the value for the *Daniell cell* (q.v.) reaction, and the theoretical voltage is obtainable from the formula $E = -\Delta G/nF$, where $nF$ is the number of coulombs transferred in the reaction; thus $E = 210 \times 10^3/2 \times 96\,490 = 1.09$ V. To obtain much higher voltages from a unit cell involves the use of highly active and unstable chemicals, as in the lithium–chlorine cell, which operates at 500 to 600 °C with an electrolyte of molten lithium chloride and gives 3.5 V.

The choice of reaction is therefore based on practical considerations: cheapness of materials, chemical stability (no deterioration when the cell is idle) and a high output-to-weight ratio; other factors may be important in special-purpose cells.

The earliest cell to be widely used was the Daniell cell. The disadvantages of employing two electrolytes are obvious, and these are avoided in the Leclanché cell, on which are based the majority of dry cells (see *dry cell*). Other cells with special uses are the *mercury cell* (q.v.), the *copper oxide cell* (q.v.), the *silver–zinc cell* (q.v.), the *zinc–air cell* (q.v.) and the *chloride cell* (q.v.), These cells all vary in the cathode reaction but all have zinc anodes, although magnesium is a promising alternative.

One basis for comparison of different cells is the cell voltage–current curve, which will normally be of the form shown in figure P.14. It starts from the theoretical maximum value, which can be calculated from the free energy change of the cell reaction. At finite currents it drops rapidly on account of an *overpotential* (q.v.) at the electrodes, and as the current is increased, the activation and concentration overpotentials increase, giving a gradually falling curve. Eventually the limiting current (see *limiting current density*) is approached, the concentration

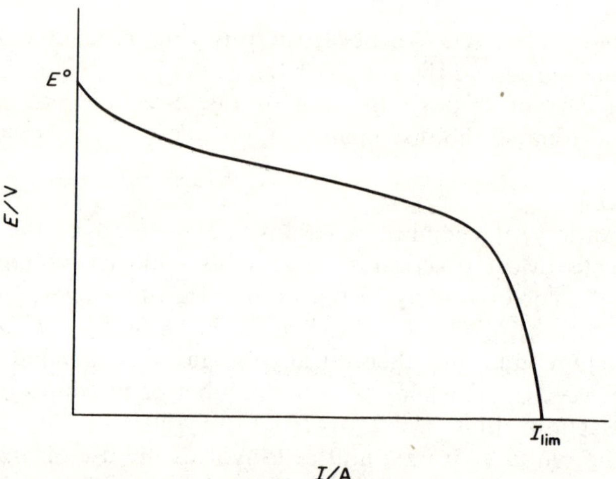


Figure P.14 Typical voltage–current curve

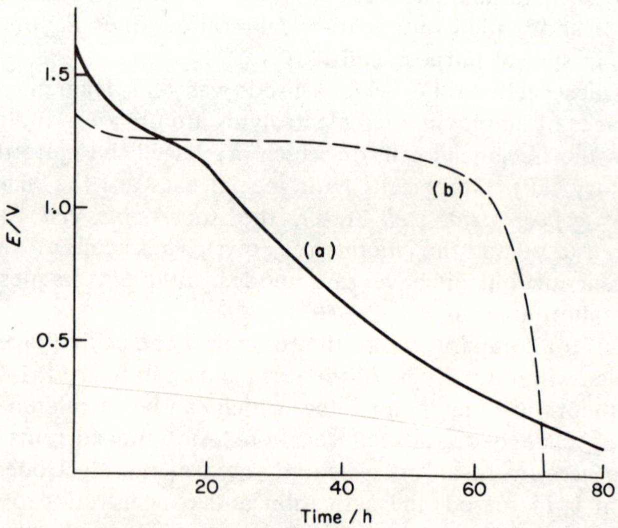


Figure P.15 Discharge curves for (a) a typical dry cell and (b) a mercury cell

overvoltage becomes very high and there is a rapid fall in the cell voltage.

Another criterion of cell performance is the voltage–time curve when the cell is discharged continuously through a fixed resistance. A series of curves for different resistances may be constructed, and also curves for the intermittent use of the cell; the latter shows how well the cell recovers, through depolarising effects and diffusion, after a period of high current withdrawal. Figure P.15 shows two typical voltage–time curves. In the mercury cell the products of electrolytic action are precipitated, and the conditions in the cell, and its voltage, remain very constant throughout its life.

# Q

**Quinhydrone electrode**
The quinhydrone electrode consists essentially of a shiny platinum electrode dipping in the test solution, which is saturated with quinhydrone (i.e. equimolar amounts of quinone, Q, and hydroquinone, $QH_2$). The quinhydrone electrode is a reversible oxidation–reduction or *redox electrode system* (q.v.):

$$Q + 2H^+ + 2e \rightleftharpoons QH_2$$

and its *electrode potential* (q.v.) can be shown to be given by

$$E(Q, QH_2) = E^{\ominus}(Q, QH_2) + \frac{RT}{2F} \ln \frac{[Q][H^+]^2}{[QH_2]}$$

The ratio $[Q]/[QH_2]$ is constant and equal to unity provided that the equilibrium is not disturbed by the presence of other oxidation–reduction systems; hence,

$$E(Q, QH_2) = E'(Q, QH_2) + \frac{RT}{F} \ln a(H^+)$$

The quinhydrone electrode therefore acts as a hydrogen ion-indicating electrode for pH values up to 8. The electrode is simply constructed by immersing bright platinum wire or foil into the test solution containing excess quinhydrone (0.5–1 g in 100 $cm^3$ of solution).

The electrode has a low electrical resistance and, in conjunction with a reference electrode, can be used with simple potentiometer circuits; it readily reaches equilibrium and can be used for micro-determinations.

It can, however, only be used in solutions of $pH<8$; above this the ratio $[Q]/[QH_2]$ no longer remains constant. Atmospheric oxidation also occurs above this pH value. The test solution is contaminated after a determination. The electrode has a serious salt error, and further it cannot be used in the presence of oxidising or reducing agents, amino compounds, ammonia and ammonium salts.

*See also* G, I & J, J & P.

# R

**Redox electrode system**

Although all electrodes are really oxidation–reduction electrodes, the term 'redox electrode' is restricted to an inert platinum electrode in contact with the oxidised and reduced forms of an electrode couple in the cell solution, e.g. $Pt \mid Fe^{3+}, Fe^{2+}$.

The standard potential of the redox system is the e.m.f. of the cell

$$\ominus \; \mathrm{Pt, H_2(g, 101\,325\ N\,m^{-2})} \left| \begin{matrix} \mathrm{HX} \\ a(\mathrm{H^+}) = 1 \end{matrix} \right\vdots \begin{matrix} \mathrm{Ox, Red} \\ a(\mathrm{ox}) = a(\mathrm{red}) \end{matrix} \left| \mathrm{Pt} \; \oplus \right.$$

in which the *liquid junction potential* (q.v.) has been eliminated. The cell reaction is

$$n/2\mathrm{H_2(g)} + \mathrm{Ox}^{z+} \rightleftharpoons n\mathrm{H^+} + \mathrm{Red}^{(z-n)+}$$

for which

$$\Delta G = -RT \ln K_{\mathrm{therm}} + RT \ln \frac{a(\mathrm{red})}{a(\mathrm{ox})} + RT \ln \frac{a^n(\mathrm{H^+})}{p^{n/2}(\mathrm{H_2})}$$

or

$$E(\mathrm{cell}) = E(\mathrm{O, R}) = E^{\ominus}(\mathrm{cell}) + \frac{RT}{nF} \ln \frac{a(\mathrm{ox})}{a(\mathrm{red})}$$

since the standard hydrogen electrode is used, where

$$E^{\ominus}(\text{cell}) = \frac{RT}{nF} \ln K_{\text{therm}} = E^{\ominus}(\text{O, R})$$

The electrode potential depends on $E^{\ominus}(\text{O, R})$ and the ratio of the activities of the oxidised and reduced forms in equilibrium. $E^{\ominus}(\text{O, R})$ is the potential of the redox electrode when $a(\text{ox}) = a(\text{red})$; it is not necessary that the activities should be unity as for other types of electrode. From a knowledge of $E^{\ominus}(\text{O, R})$ the potential of any mixture of oxidised and reduced forms can be calculated (figure R.1).

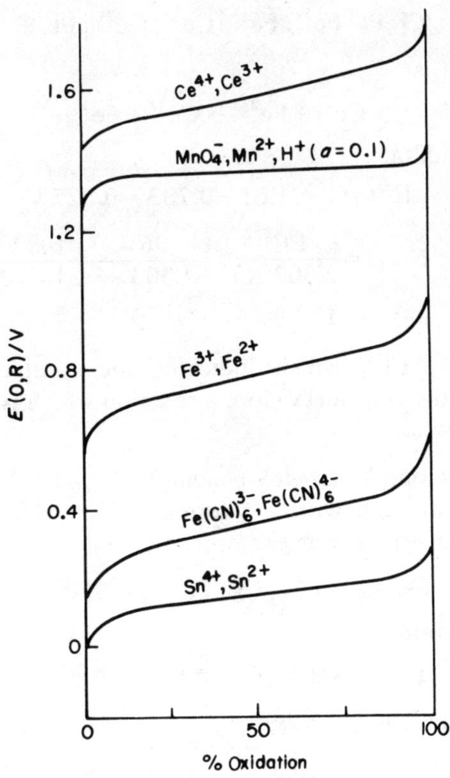


Figure R.1 Oxidation–reduction potentials

*Redox electrode system*

The redox potential is the potential adopted by the platinum electrode due to the equilibrium

$$\text{oxidised state} + ne \rightleftharpoons \text{reduced state}$$

If the equilibrium lies to the left (right), the inert electrode adopts a negative (positive) potential and the redox system is a good reducing (oxidising) agent. The more positive the redox potential of a system, the better oxidising agent is that system; from which it follows that a system with a more positive potential will oxidise one with a less positive potential.

Thus, for the cell

$$\ominus\ \text{Pt} \mid Fe^{3+}, Fe^{2+} \,\vdots\, Ce^{4+}, Ce^{3+} \mid \text{Pt}\ \oplus$$

the cell reaction is

$$Ce^{4+} + Fe^{2+} \rightarrow Ce^{3+} + Fe^{3+}$$

From tabulated data,

$$E^{\ominus}(\text{cell}) = 1.61 - 0.783 = 0.827\ \text{V}$$

whence

$$\log K_{\text{therm}} = \frac{FE^{\ominus}}{2.303\ RT} = \frac{96\,487 \times 0.827}{2.303 \times 8.314 \times 298}$$

$$= 13.98$$

i.e. $K_{\text{therm}} = 9.65 \times 10^{13}$, which indicates the complete oxidation of iron(II) salts by the cerium(IV) ion, a reaction which is thus suitable for analytical purposes.

*Determination of standard redox potentials*

(1) Using a simple cell with a reference electrode, as for a normal *electrode potential* (q.v.). For example, the cell

$$\ominus\ \text{Ag, AgCl} \mid Cl^-, Fe^{2+}, Fe^{3+} \mid \text{Pt}\ \oplus$$

has the cell reaction

$$Fe^{3+} + \text{Ag(s)} + Cl^- \rightleftharpoons Fe^{2+} + \text{AgCl(s)}$$

for which the e.m.f. is given by

$$E(\text{cell}) - \frac{RT}{F} \ln \frac{c(Fe^{3+})c(Cl^-)}{c(Fe^{2+})} = E^{\ominus}(\text{cell}) + \frac{RT}{F} \ln \frac{\gamma(Fe^{3+})\gamma(Cl^-)}{\gamma(Fe^{2+})}$$

From a knowledge of $E$(cell) at different concentrations of the oxidised and reduced forms, the graph of

$$E(\text{cell}) - \frac{RT}{F} \ln \frac{c(\text{Fe}^{3+})c(\text{Cl}^-)}{c(\text{Fe}^{2+})}$$

against $I^{1/2}$ is linear and of intercept

$$E^\ominus(\text{cell}) = E^\ominus(\text{Fe}^{3+}, \text{Fe}^{2+}) - E^\ominus(\text{AgCl, Ag, Cl}^-)$$

This is not a very good method, since exact knowledge of the concentrations of the oxidised and reduced forms is difficult to obtain.

(2) From equilibrium constant measurements; this is probably the most accurate method. The cell reaction of the hypothetical cell

$$\text{Pt} \,|\, \text{Fe}^{2+}, \text{Fe}^{3+} \,\vdots\, \text{Ag}^+ \,|\, \text{Ag}$$

is

$$\text{Fe}^{2+} + \text{Ag}^+ \rightleftharpoons \text{Fe}^{3+} + \text{Ag(s)}$$

and the standard e.m.f. is given by

$$E^\ominus(\text{cell}) = \frac{RT}{F} \ln K_{\text{therm}} = \frac{RT}{F} \ln \left( \frac{a(\text{Fe}^{3+})}{a(\text{Fe}^{2+})a(\text{Ag}^+)} \right)_e$$

$$= E^\ominus(\text{Ag}^+, \text{Ag}) - E^\ominus(\text{Fe}^{3+}, \text{Fe}^{2+})$$

The *equilibrium constant*‡, $K_c$, is calculated from measurements of the equilibrium concentrations of iron(II), iron(III) and silver ions when a solution of iron(III) perchlorate containing excess perchloric acid (to prevent hydrolysis) is shaken with finely divided metallic silver at different ionic strengths.

$$K_{\text{therm}} = K_c \times \frac{\gamma(\text{Fe}^{3+})}{\gamma(\text{Fe}^{2+})\gamma(\text{Ag}^+)}$$

Since

$$\log \gamma_i = -Az_i^2 I^{1/2} + CI$$

it follows that

$$\log K_{\text{therm}} = \log K_c - 4AI^{1/2} + CI$$

Thus from measured values of $K_c$, $E^\ominus$(cell) and, hence, $E^\ominus$(O, R) for the redox system can be calculated.

(3) From *potentiometric titration* (q.v.) curves; this method is mainly used for systems involving organic and biological compounds. Only

approximate values of the standard redox potential are obtained since concentrations are used in place of activities, and the value applies only to a specified pH and ionic strength; this is sometimes known as the 'formal electrode potential', $E^{\ominus'}$.

The pure oxidised (reduced) form of the substance, dissolved in a buffer solution of known pH and *I*, is titrated against a reducing (oxidising) agent in the complete absence of air (figure R.2). After each addition of titrant, the e.m.f. of the cell, comprising a platinum and reference electrode, is measured and plotted as a titration curve (figure R.3):

$$\ominus \ \text{Hg, Hg}_2\text{Cl}_2 \left| \begin{array}{c}\text{saturated}\\ \text{KCl aq.}\end{array} \right\vdots \text{Red, Ox} \,\big|\, \text{Pt} \ \oplus$$

for which

$$E(\text{cell}) = E^{\ominus}(\text{O, R}) - E(\text{cal}) + \frac{RT}{nF} \ln \frac{a(\text{ox})}{a(\text{red})}$$

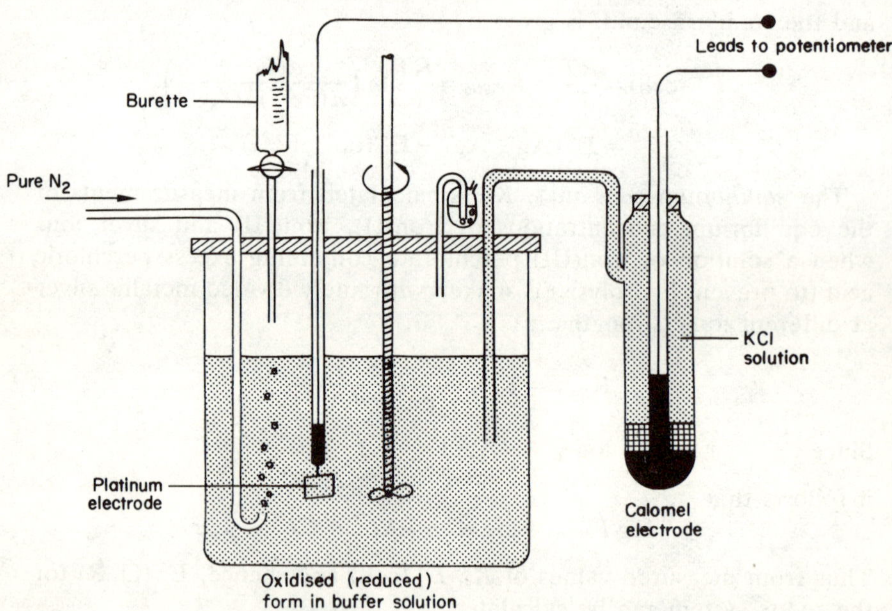


Figure R.2 Experimental arrangement for measuring the electrode potential of an oxidation–reduction electrode during potentiometric titration

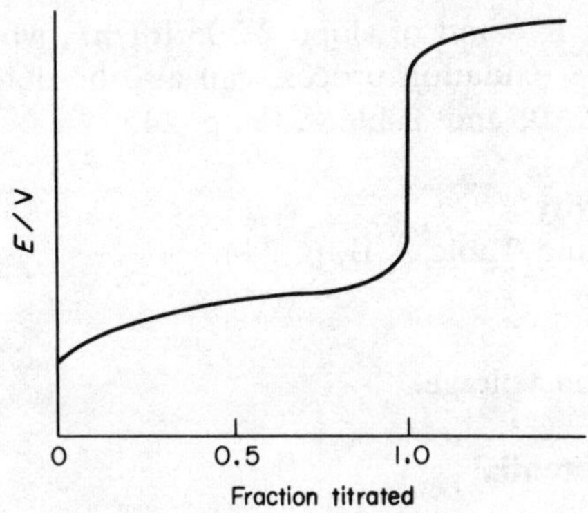


Figure R.3 Variation of e.m.f. of cell during titration

From the volume of the titrant (oxidising agent), $t$, added at different times to the reduced form, and the volume required for rapid change, $T$, i.e. 100% oxidised, it follows that $c(\text{ox})$ is proportional to $t$ and $c(\text{red})$ to $(T-t)$; whence,

$$E(\text{cell})+E(\text{cal}) = E^{\ominus'}(\text{O, R})+\frac{RT}{nF}\ln\frac{t}{(T-t)}$$

where $E^{\ominus'}(\text{O, R})$ is the formal electrode potential at the given pH and $I$. The graph of $E(\text{cell})+E(\text{cal})$ against $\log t/(T-t)$ is linear (figure

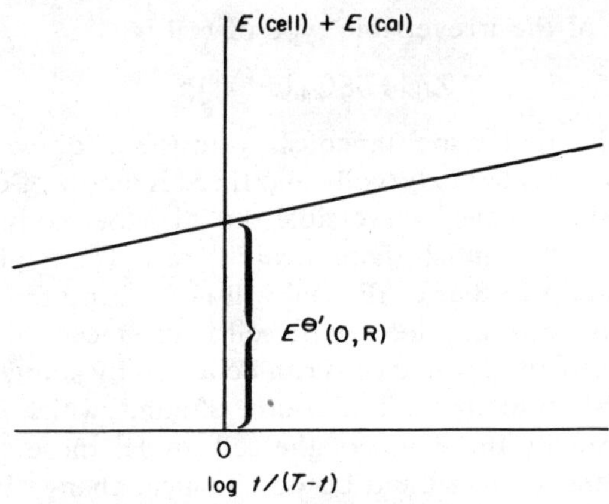


Figure R.4 Graph of electrode potential during titration against $\log t/(T-t)$

R.4), of intercept $E^{\ominus'}$ and of slope 2.303 $RT/nF$, whence the number of electrons in the oxidation process can also be determined.

*See also* G, J & P; and Table A.IV, p. 245.

**Relative permittivity**
*See* Permittivity and Table A.II, p. 244.

**Residual current**
*See* Decomposition voltage.

**Resistance overpotential**
*See* Overpotential.

**Reversible galvanic cell**
Any arrangement, consisting of two electrodes and an electrolyte capable of undergoing spontaneous chemical reaction to produce an electric current when the electrodes are joined externally, is called a galvanic cell.

Cells of this type may be divided into two categories, depending on whether a chemical reaction occurs even before there is a flow of current (irreversible cells) or whether there is no reaction until the electrodes are joined externally and there is a flow of current (reversible cells).

An example of the irreversible type of cell is

$$\mathrm{Zn} \mid \mathrm{H_2SO_4}\ \text{(dil.)} \mid \mathrm{Pt}$$

in which the zinc reacts spontaneously with the acid, even before the electrodes are connected externally and there is a flow of current. Cells of this sort are always irreversible in the thermodynamic sense and, hence, are not of much theoretical interest. The application of an e.m.f. in opposition to that of the cell will not restore the system to its original state; instead, an electrolytic cell is produced.

The reversibility of galvanic cells can be tested by connecting the cell under consideration to an external source of e.m.f. which is adjusted so as to balance exactly the e.m.f. of the cell; under these conditions no current flows, and there should be no chemical change in the cell. If the external e.m.f. is decreased by an infinitesimal amount, current will flow from the cell and a chemical change, proportional to the quantity of electricity passed, should occur. On the other hand, if the opposing e.m.f. is slightly greater than that of the cell, current will flow in the opposite direction and the cell reaction should be reversed. Both electrode reactions of a reversible cell must be reversible; i.e. both electrodes must be reversible.

Such electrochemical cells will only behave reversibly when the current flowing is infinitesimally small and the system is virtually always in equilibrium. If large currents flow, concentration gradients are set up and the cell can no longer be regarded as in a state of equilibrium. Care must therefore be taken when measuring the e.m.f. of a reversible cell using a *potentiometer* (q.v.).

An example of a reversible cell (figure R.5) is:

$$\ominus \ Zn \,|\, ZnCl_2, aq \,|\, Cl_2, Pt \ \oplus; \qquad E^{\ominus}(\text{cell}) = 2.12\ V$$

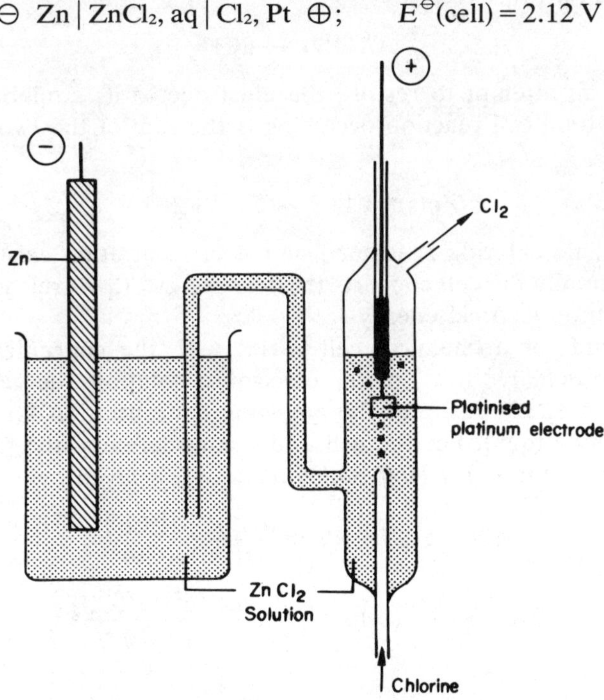


Figure R.5 Typical zinc–chlorine cell

in which the zinc is in equilibrium with zinc ions in solution, and the gaseous chlorine with the chloride ions. Experimentally the potential of the zinc electrode is negative with respect to that of the chlorine electrode. Connection of the electrodes externally (e.g. through a potentiometer) results in an electron flow from the zinc to the chlorine electrode through this circuit; this disturbs the electrode equilibrium and more electrons are required to try to maintain the zinc electrode at its equilibrium potential. These are provided by the ionisation of more zinc (oxidation):

$$Zn \rightarrow Zn^{2+} + 2e$$

As the electrons arrive at the chlorine electrode, through the external circuit, this becomes more negative than the equilibrium potential, and the reduction process

$$Cl_2 + 2e \rightarrow 2Cl^-$$

occurs in an attempt to restore the electrode to its equilibrium position. The total cell reaction occurring is the sum of the two electrode reactions

$$Zn(s) + Cl_2(g) \rightarrow Zn^{2+} + 2Cl^-$$

in which zinc chloride is formed at the concentration existing in the cell. Essentially the cell converts the *free energy*‡ (q.v.) released by this reaction into electrical energy.

The e.m.f. of a chemical cell varies with the concentration and, hence, the *activity*‡ (q.v.) of the electrolyte solution, the gas pressure for gaseous electrodes, and the amalgam concentration for an *amalgam electrode* (q.v.). For the cell under consideration, the free energy change, from the *van't Hoff isotherm*‡ (q.v.), is given by

$$\Delta G = \Delta G^{\ominus} + RT \ln \frac{a(Zn^{2+})a^2(Cl^-)}{a(Zn)p(Cl_2)}$$

or

$$E(\text{cell}) = E^{\ominus}(\text{cell}) - \frac{RT}{2F} \ln \frac{a(Zn^{2+})a^2(Cl^-)}{p(Cl_2)}$$

where

$$E^{\ominus}(\text{cell}) = -\frac{\Delta G^{\ominus}}{nF} = \frac{RT}{nF} \ln K_{\text{therm}}$$

If $p(Cl_2)$ is kept constant at atmospheric pressure, and $m$ is the molality of zinc chloride in solution, then

$$E(\text{cell}) = E^{\ominus}(\text{cell}) - \frac{RT}{2F}\ln 4m^3 - \frac{3RT}{2F}\ln \gamma_{\pm}$$

from measurements of $E(\text{cell})$ at different (small) values of $m$, the graph of $\left[E(\text{cell}) + \frac{RT}{2F}\ln 4m^3\right]$ against $I^{1/2}$ will be linear and of intercept $E^{\ominus}(\text{cell})$; hence, $\gamma_{\pm}$ can be calculated.

Another widely used simple galvanic cell consists of the hydrogen and silver–silver chloride electrodes in combination:

$$\ominus\ \text{Pt}, H_2(g) \mid H^+\,Cl^- \mid \text{AgCl}, \text{Ag}\ \oplus$$

for which the electrode and cell reactions are

$$\tfrac{1}{2}H_2(g) \rightarrow H^+ + e$$

$$\text{AgCl(s)} + e \rightarrow \text{Ag(s)} + Cl^-$$

$$\tfrac{1}{2}H_2(g) + \text{AgCl(s)} \rightarrow \text{Ag(s)} + \underbrace{H^+ + Cl^-}_{(m)}$$

For this cell, as written, $E$ is positive and, hence, the forward reaction is spontaneous. The variation of e.m.f. of this cell with concentration may be obtained by an alternative method using the *Nernst equation* (q.v.) for electrode potentials:

$$\begin{aligned}E(\text{cell}) &= E(\text{AgCl, Ag, Cl}^-) - E(\text{H}^+, \text{H}_2)\\ &= E^{\ominus}(\text{AgCl, Ag, Cl}^-) - \frac{RT}{F}\ln a(\text{Cl}^-)\\ &\qquad - E^{\ominus}(\text{H}^+, \text{H}_2) - \frac{RT}{F}\ln\frac{a(\text{H}^+)}{p^{1/2}(\text{H}_2)}\\ &= E^{\ominus}(\text{cell}) - \frac{RT}{F}\ln a(\text{H}^+)a(\text{Cl}^-) - \frac{RT}{F}\ln p^{1/2}(\text{H}_2)\end{aligned}$$

When $p(H_2)$ is kept constant at atmospheric pressure,

$$E(\text{cell}) = E^{\ominus}(\text{cell}) - \frac{2RT}{F}\ln a_{\pm} = E^{\ominus}(\text{cell}) - \frac{2RT}{F}\ln m_{\pm} - \frac{2RT}{F}\ln \gamma_{\pm}$$

thus the graph of $\left[E(\text{cell}) + \frac{2RT}{F} \ln m_{\pm}\right]$ against $I^{1/2}$ is linear and of intercept $E^{\ominus}(\text{cell})$; and this is equal to $E^{\ominus}(\text{AgCl, Ag, Cl}^-)$, since for the hydrogen electrode $E^{\ominus}(\text{H}^+, \text{H}_2) = 0$.

The general equation for the e.m.f. of reversible cells is

$$E(\text{cell}) = E^{\ominus}(\text{cell}) - \frac{RT}{nF} \ln \frac{\Pi a(\text{products})}{\Pi a(\text{reactants})}$$

where $n$ is the number of electrons involved in the process.

Studies of the e.m.f. of reversible galvanic cells permit the determination of the value of a standard *electrode potential* (q.v.), an *equilibrium constant*‡ (q.v.), *activity*‡ (q.v.) and *activity coefficient*‡ (q.v.).

*See also* De, G.

# S

**Sacrificial protection**

*See* Corrosion.

**Salt bridge**

A salt bridge is used to connect a reference electrode with the test solution or to connect two solutions such that the *liquid junction potential* (q.v.) is reduced as far as possible. A salt bridge consists of an inverted U-tube plugged at the ends either with sintered discs of fine porosity or with tightly rolled spirals of filter paper or asbestos. The tube is filled with the required bridge solution and the bridge stored with the legs dipping in vessels containing the same solution. Bridges of this type have the advantage over agar-salt bridges that they last indefinitely and do not suffer from syneresis or shrinkage of the gel. Where contamination of the test solution with the bridge solution must be kept to a minimum, an agar bridge must be used.

A saturated potassium chloride agar bridge is made by adding 3–5 g of powdered agar, in small portions at a time, so that the solution does not froth and boil over, to 100 cm$^3$ of a saturated solution of potassium chloride at 100 °C on a steam-bath. The solution is kept at 100 °C until all the agar has passed into solution and then 10–15 g of solid potassium chloride are added to produce excess of the solid.

Other agar-salt bridges can be made by dissolving 3–5 g of powdered agar in 100 $cm^3$ of potassium or ammonium nitrate solution (2 mol $dm^{-3}$); this type of bridge is commonly used to minimise junction potentials. Such a bridge is of use with halide solutions during titration with silver nitrate.

**Sedimentation potential**
*See* Dorn effect.

**Sign convention**
The IUPAC or Stockholm convention of e.m.f. values and electrode potentials in reversible galvanic cells, now widely adopted, is used throughout this book. The following is a summary of the conventions relating to galvanic cells and electrode potentials.

(1) The process of reduction involves a gain of electrons and oxidation a loss of electrons,

$$\text{oxidised state} + ne \underset{\text{oxidation}}{\overset{\text{reduction}}{\rightleftharpoons}} \text{reduced state}$$

e.g.

$$Cu^{2+} + 2e \rightleftharpoons Cu$$
$$Fe^{3+} + e \rightleftharpoons Fe^{2+}$$

(2) The sign of the electrode potential, on open circuit, is that which the electrode has with respect to the solution, i.e. for the electrode $Cu^{2+} | Cu$, if there is a tendency for $Cu^{2+}$ to be discharged, then the electrode will become positively charged with respect to the solution, so the electrode potential is positive; this corresponds to the reaction

$$Cu^{2+} + 2e \rightarrow Cu$$

Thus a positive potential corresponds to a reduction process; electrode potentials on this convention are thus really reduction potentials.

(3) The hydrogen electrode, with hydrogen gas at unit *fugacity*‡ (1 atmosphere) in a solution of hydrogen ions ($a(H^+) = 1$) is arbitrarily taken as the zero of electrode potentials at all temperatures, i.e. for the half-cell

$$H^+[a(H^+) = 1] \,|\, H_2(g, 101\,325\ N\ m^{-2}), Pt; \qquad E^{\ominus}(H^+, H_2) = 0$$

*Sign convention*

(4) The half-cells representing the reduction electrode potentials on open circuit written

$$\left.\begin{array}{l} Zn^{2+} \mid Zn \\ Cl^- \mid Cl_2, Pt \\ Cl^- \mid AgCl, Ag \\ Fe^{2+}, Fe^{3+} \mid Pt \end{array}\right\} \begin{array}{l}\text{imply that}\\ \text{the electrode}\\ \text{processes are}\\ \text{reduction}\end{array} \left\{\begin{array}{l} Zn^{2+} + 2e \rightarrow Zn(s) \\ \frac{1}{2}Cl_2(g) + e \rightarrow Cl^- \\ AgCl(s) + e \rightarrow Ag(s) + Cl^- \\ Fe^{3+} + e \rightarrow Fe^{2+} \end{array}\right.$$

and further that the potentials of these electrodes are those of the type (in which the junction potential is ignored):

$$Pt, H_2(g, 101\,325\ N\ m^{-2}) \mid H^+[a(H^+) = 1] \vdots X^{n\pm} \mid X$$

where

$$\begin{aligned} E(\text{cell}) &= E(\text{RH electrode}) - E(\text{LH electrode}) \\ &= E(X^{n\pm}, X) - E(H^+, H_2) = E(X^{n\pm}, X) - 0 \end{aligned}$$

i.e.

$$\left.\begin{array}{l} Pt, H_2 \mid H^+ \vdots Zn^{2+} \mid Zn \\ Pt, H_2 \mid H^+ \vdots Cl^- \mid Cl_2, Pt \\ Pt, H_2 \mid H^+ \vdots Cl^- \mid AgCl, Ag \\ Pt, H_2 \mid H^+ \vdots Fe^{2+}, Fe^{3+} \mid Pt \end{array}\right\} \begin{array}{c}\text{implying}\\ \text{the}\\ \text{reactions}\end{array}$$

$$\begin{array}{ll} & E^{\ominus}(\text{cell})/V \\ \left\{\begin{array}{l} \frac{1}{2}H_2 + \frac{1}{2}Zn^{2+} \rightarrow \frac{1}{2}Zn(s) + H^+ \\ \frac{1}{2}H_2 + \frac{1}{2}Cl_2 \rightarrow H^+ + Cl^- \\ \frac{1}{2}H_2 + AgCl(s) \rightarrow Ag(s) + H^+ + Cl^- \\ \frac{1}{2}H_2 + Fe^{3+} \rightarrow Fe^{2+} + H^+ \end{array}\right. & \begin{array}{r} -0.761 \\ 1.358 \\ 0.2225 \\ 0.783 \end{array} \end{array}$$

(5) The sign of $E$(cell) is the sign of the right-hand electrode. If $E$(cell) is positive, then (a) the process at the right-hand electrode is reduction; (b) the over-all cell process is spontaneous from left to right, since $-\Delta G = nFE$, when $E > 0$, $\Delta G < 0$; (c) positive ions move from left to right through the cell, while negative ions move in the reverse direction, and electrons pass from left to right through the external circuit.

In the first of the above cells, comprising the standard hydrogen and zinc electrodes, the spontaneous cell reaction is, in fact, the reverse; so, to conform with the experimental fact that the hydrogen electrode is more positive than the zinc, the cell should be written

$$\ominus \ Zn \mid Zn^{2+} \vdots H^+ \mid H_2(g), Pt \ \oplus$$

for which $E^\ominus(\text{cell}) = 0.761$ V; hence, $E^\ominus(Zn^{2+}, Zn) = -0.761$ V; whence the spontaneous electrode and the cell reactions are

$$\begin{array}{c} \tfrac{1}{2}Zn(s) \rightarrow \tfrac{1}{2}Zn^{2+} + e \\ H^+ + e \rightarrow \tfrac{1}{2}H_2(g) \\ \hline H^+ + \tfrac{1}{2}Zn(s) \rightarrow \tfrac{1}{2}Zn^{2+} + \tfrac{1}{2}H_2(g) \end{array}$$

The positive cell e.m.f. indicates the forward reaction is spontaneous and conforms to the stated conditions.

For the reaction

$$\tfrac{1}{2}H_2(g) + AgI(s) \rightarrow Ag(s) + HI$$

$\Delta G$ changes sign when the molality of HI is about 0.1 mol kg$^{-1}$, so that the cell

$$H_2(g), Pt \mid HI\,(m) \mid AgI(s), Ag$$

changes its polarity at about this concentration. The e.m.f. of the cell (as written) is positive for $m < 0.1$ and negative for $m > 0.1$ mol kg$^{-1}$.

(6) The equation representing the *electrode potential* (q.v.) is

$$E(O, R) = E^\ominus(O, R) + \frac{RT}{nF}\ln\frac{a(\text{ox})}{a(\text{red})}$$

The equation for the e.m.f. of a chemical or *reversible galvanic cell* (q.v.) is

$$\begin{aligned} E(\text{cell}) &= E(\text{RH}) - E(\text{LH}) = E^\ominus(\text{RH}) - E^\ominus(\text{LH}) - \frac{RT}{nF}\ln\frac{\Pi a(\text{products})}{\Pi a(\text{reactants})} \\ &= \frac{RT}{nF}\ln K_{\text{therm}} - \frac{RT}{nF}\ln\frac{\Pi a(\text{products})}{\Pi a(\text{reactants})} \end{aligned}$$

For a cell in which the standard hydrogen electrode is one electrode, the e.m.f. is given by

$$E(\text{cell}) = E^\ominus(\text{cell}) - \frac{RT}{nF}\ln\frac{a(\text{ox})}{a(\text{red})}$$

(7) When a *liquid junction potential* (q.v.) is included, $E_j$ is added when $E$(cell) is positive and subtracted when $E$(cell) is negative.

*See also* De, I & J.

**Silver coulometer**

Either the deposition of metallic silver on an inert cathode or the dissolution of silver from a silver anode may be used for measuring the quantity of electricity passed through a circuit. The commoner type of silver coulometer depends upon the deposition of metallic silver on an inert cathode (figure S.1). This coulometer consists of a sintered glass crucible (A) of No. 3 porosity, supported on a glass frame (B) so that there is free movement of the solution through the sintered disc. The cathode consists of a piece of platinum foil (1 $cm^2$ area) welded onto platinum wire, while the anode is a coil of stout silver wire. The electrodes and crucible are cleaned and washed. The cathode and crucible are dried together in an oven at 150 °C to constant weight. The electrodes are then connected in the circuit and the beaker filled with 15% silver nitrate solution (the silver nitrate must be recrystallised and free from organic material). The current, which should not exceed 10 mA $cm^{-2}$ of cathode surface, is passed through the solution; silver will be deposited on the cathode, and any small fragments which break away will be collected in the crucible. After switching off the current, the platinum cathode is disconnected and rested in the crucible. The crucible and electrode are removed from the beaker, washed and dried to constant weight. The accuracy of this method is limited by

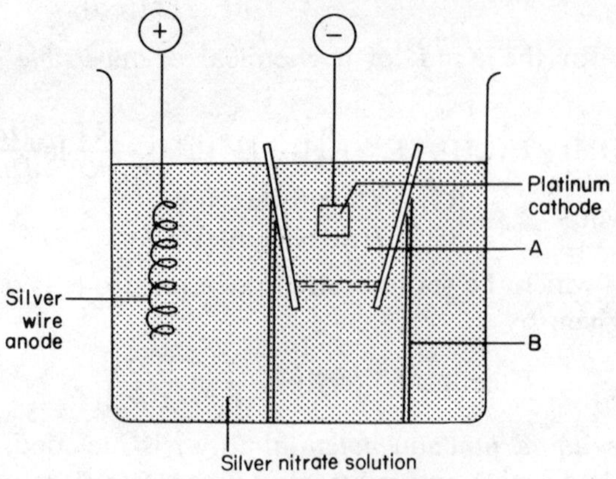


Figure S.1 Silver coulometer

the care of manipulation and the accuracy of weighing (0.001 118 g of silver is deposited by 1 coulomb).

Although this coulometer gives very accurate results, there is the possibility of some occlusion of water or nitrate in the silver deposit. For this reason, the dissolution reaction was preferred in the most recent determination of the *Faraday constant* (q.v.).

**Silver electrode**

The silver electrode, reversible to the silver ion in solution, has an *electrode potential* (q.v.) given by

$$E(\mathrm{Ag^+, Ag}) = E^{\ominus}(\mathrm{Ag^+, Ag}) + \frac{RT}{F}\ln a(\mathrm{Ag^+})$$

The standard electrode potential, $E^{\ominus}(\mathrm{Ag^+, Ag}) = 0.7991$ V.

For accurate work, the silver electrode is coated with a fresh film of silver. A current is passed through a solution of $KAg(CN)_2$ (freshly prepared from silver nitrate and potassium cyanide) between the electrode to be plated, which is the cathode, and a silver anode. An electrode prepared from 22 S.W.G. silver wire is adequate for most purposes.

*See also* G, I & J, J & P.

**Silver–silver chloride electrode**

The silver–silver chloride electrode, used as a reference electrode, consists of a strip or disc of silver, on which is deposited a film of silver chloride. It behaves as a reversible chlorine electrode, with a potential given by

$$E(\mathrm{AgCl, Ag, Cl^-}) = E^{\ominus}(\mathrm{AgCl, Ag, Cl^-}) - \frac{RT}{F}\ln a(\mathrm{Cl^-})$$

The standard electrode potential is also given by

$$E^{\ominus}(\mathrm{AgCl, Ag, Cl^-}) = E^{\ominus}(\mathrm{Ag^+, Ag}) + \frac{RT}{F}\ln K_s(\mathrm{AgCl})$$

The electrode is prepared by immersing a freshly plated *silver electrode* (q.v.) in hydrochloric acid (0.1 mol dm$^{-3}$) and connecting to the positive pole of an accumulator. Using a platinum electrode as cathode, a current of 2.5 mA $cm^{-2}$ over the whole surface is passed for 30 min. An even purple deposit of silver chloride will be formed; this is unaffected by sunlight. The electrode potential is reproducible to ±0.02 mV.

The standard electrode potential at $T$ K is given by

$$E^{\ominus}(\mathrm{AgCl, Ag, Cl^-}) = 0.222\,39 - 645.52\times10^{-6}(T-298) - 3.284\times10^{-6}(T-298)^2 + 9.948\times10^{-9}(T-298)^3$$

Silver–silver bromide and silver–silver iodide electrodes prepared in a similar manner behave as reversible bromine and iodine electrodes, respectively.

*See also* G, I & J.

**Silver–zinc cell**

Cells of this type resemble the *mercury cell* (q.v.) in their chemistry, and have similar uses as miniature cells. The cathode reaction in the usual cell is

$$Ag_2O + H_2O + 2e \rightarrow 2Ag + 2OH^-$$

but cells can be prepared with silver peroxide cathodes (prepared electrochemically) which react:

$$AgO + H_2O + 2e \rightarrow Ag + 2OH^-$$

The cells are more costly than mercury cells, but have a higher energy–volume ratio, and give higher voltages ($Ag_2O$, 1.58 V; AgO, 1.86 V).

The reactions of silver–zinc cells are reversible, and they can be made up as storage batteries. Their output–weight ratio is far higher than that of a lead accumulator, and they give a stable voltage under discharge. Besides their expense, however, they have the disadvantage of tolerating only a limited number of chargings, owing to a deterioration in the physical condition of the zinc anode. They have been used mainly in aircraft and aerospace applications. Cadmium may replace zinc; the silver–cadmium accumulator has a lower voltage but a considerably longer life.

*See also* Mi, P.

**Sodium, electrometallurgy**

It is possible to discharge sodium electrolytically from an aqueous solution if a mercury cathode is used. On the basis of the standard electrode potentials [$E^{\ominus}(Na^+, Na) = -2.71$ V], the decomposition of water to form hydrogen at the cathode would be a far easier process even when allowance is made for the high hydrogen *overpotential* (q.v.) at mercury. However, the fact that sodium forms intermetallic compounds with mercury which are soluble in mercury and diffuse away so reduces the activity of the sodium at the cathode surface, and its tendency to re-ionise that the discharge of sodium becomes the preferred process. This method is not used commercially because of the high cost of extraction of sodium from its amalgam, and recourse is had to fused electrolytes.

Metallic sodium melts at 370.5 K and boils at 1153 K. A low working temperature is therefore desirable. Sodium hydroxide melts at 591 K and sodium chloride at 1073 K, and in the earlier process, now seldom used, sodium hydroxide was employed at a temperature just above its melting point. The primary electrode reactions are:

Cathode: $Na^+ + e \rightarrow Na$

Anode: $4OH^- \rightarrow 2H_2O + O_2 + 4e$

The water formed is readily electrolysed to give hydrogen at the cathode, and the current efficiency for sodium formation cannot exceed 0.5. It is further reduced by oxidation of the sodium discharged, which is fairly soluble in the melt. Sodium chloride is now preferred as the electrolyte.

One of the difficulties in the use of molten electrolytes is the solubility of the metal in the electrolyte, and the formation of metallic clouds consisting of a colloidal dispersion of the metal and one of its compounds. The resulting losses increase with rise of temperature, and at 1073 K the efficiency of sodium production would be exceedingly low. The temperature of the melt is therefore reduced by using a mixture of sodium chloride and calcium chloride (or sodium carbonate) which is molten at about 873 K. The standard electrode potential of calcium, $E^{\ominus}(Ca^{2+}, Ca) = -2.87$ V, is slightly more negative than that of sodium; only about 1% calcium is produced at the cathode, and most of this separates when the sodium solidifies.

The electrolysis is carried out in a firebrick-lined steel vessel with a central graphite anode, from which the discharged chlorine rises and is carried away. This is surrounded by an annular iron cathode. The molten sodium produced here is lighter than the electrolyte and rises into the collecting space from which it travels into a cooling vessel. The decomposition voltage is about 3.3 V, but in practice a voltage about double this value is used, and the temperature is maintained at 873 K. The current efficiency is about 80%, and, hence, the power consumption is no greater than for the sodium hydroxide process.

*See also* Mi, P.

**Solions**

Solions are small electrolytic cells devised to act as electrical control elements. A closed cell is used containing any suitable reversible system such as a very dilute solution of iodine in potassium iodide. A constant voltage is applied between inert electrodes, when iodine is reduced at the cathode while iodine is oxidised at the anode:

$$I_3^- + 2e \rightarrow 3I^-$$

and

$$3I^- \rightarrow I_3^- + 2e$$

the net change being nil. The voltage applied is sufficient to ensure that the limiting current density is attained. If now the temperature rises, diffusion rates are increased, and the increase in current measures the increase in temperature. Other effects, such as pressure changes, can be monitored in the same way by adaptations of the cell.

**Solvent correction**

The solvent used in conductance measurements will usually have a very small but measurable conducting power which must be allowed for in deriving the correct conductance of the dissolved electrolyte. The conductance of the solvent will be partly due to its own ionisation and partly to dissolved impurities, and it should obviously be made as small as possible. For aqueous solutions, the water obtained from a well-washed 'mixed-bed' ion-exchange column should have a conductivity of $1\text{–}2 \times 10^{-7}\ \Omega^{-1}\ \text{cm}^{-1}$ at 25 °C, attributable mainly to the $H^+$ and $OH^-$ ions of water itself, and with water of this quality it is only in very dilute solutions that a solvent correction is necessary. If an open cell has to be used, however, such water would rapidly deteriorate on exposure to the air and its conductance would increase. It is then better to use water that is in equilibrium with the air; this will have a conductivity of about $8\text{–}10\times10^{-7}\ \Omega^{-1}\ \text{cm}^{-1}$, due mainly to dissolved carbon dioxide, and this should not alter greatly during the measurements.

In either case, the normal solvent correction is a straightforward subtraction:

$$\kappa(\text{solution}) - \kappa(\text{solvent}) = \kappa(\text{solute})$$

or, for measurements made in the same cell,

$$1/R(\text{solution}) - 1/R(\text{solvent}) = 1/\text{R}(\text{solute})$$

Exceptionally, the normal solvent correction does not apply. In an acid solution the ionisations

$$H_2O \rightleftharpoons H^+ + OH^-$$

and

$$H_2CO_3 \rightleftharpoons H^+ + HCO_3^-$$

will be completely suppressed by the excess of the common ion $H^+$ from the acid solute, and no solvent correction should be deducted. In alkaline solutions and solutions of a bicarbonate or carbonate the ionisation of any $CO_2$ present will be affected, so for 'equilibrium water' the true solvent correction will require calculation.

**Standard cell**

*See* Clark cell; Weston cell.

**Standard electrode potential**

*See* Electrode potential.

**Standards, conductance**

The standards universally adopted are those of Grinnell Jones and Bradshaw (*J. Amer. Chem. Soc.*, 1933, **55,** 1780). These are for aqueous KCl solutions of specified concentration. They are based on the value for the conductivity of pure mercury, used in defining the

Table S.1 Conductivity values of KCl solutions used as standards

| Solution | Concentration | $\kappa/\Omega^{-1}\,\mathrm{cm}^{-1}$ at 0 °C | 18 °C | 25 °C |
|---|---|---|---|---|
| (a) | 1.0 D | $0.065\,17_6$ | $0.097\,83_8$ | $0.111\,34_2$ |
| (b) | 0.1 D | $0.007\,137_9$ | $0.011\,166_7$ | $0.012\,856_0$ |
| (c) | 0.01 D | $0.000\,773\,6_4$ | $0.001\,220\,5_2$ | $0.001\,408\,7_7$ |

international ohm, and were obtained from this by a series of comparisons, (a) 71.1352 g KCl per 1000 g solution (1 'demal'), (b) 7.419 13 g KCl per 1000 g solution (0.1 D), (c) 0.745 263 g KCl per 1000 g solution (0.01 D), the conductivity values of which are listed in table S.1.

In every case the weights are the corrected values for weights in vacuum, and a *solvent correction* (q.v.) must be applied to the measured conductance. The KCl is recrystallised several times and then heated to fusion in a platinum dish; it must give no test for alkali with phenolphthalein.

Even solution (c) may be too highly conducting to be suitable for a cell designed for very dilute solutions. In this case, calibration is based on interpolation from the very accurate measurements of Shedlovsky on more dilute aqueous KCl solutions at 25 °C. Up to a concentration $c/\mathrm{mol\,dm^{-3}} = 0.001$ these are represented by the equation

$$\kappa = 0.149\,92\,c - 0.094\,67\,c^{3/2}$$

*See also* R & S.

**Streaming potential**

The streaming potential effect is the reverse of *electro-osmosis* (q.v.); when liquid flows through a capillary tube or through a plug of finely divided material, a potential difference arises between the two ends of the material (figure S.2). The hydrostatic pressure is measured by the difference of level maintained by the reservoirs, and a plug of the material under study is contained between the two platinum gauze electrodes, which are connected to a valve potentiometer.

The effect is a consequence of the *electrical double layer* (q.v.), and is

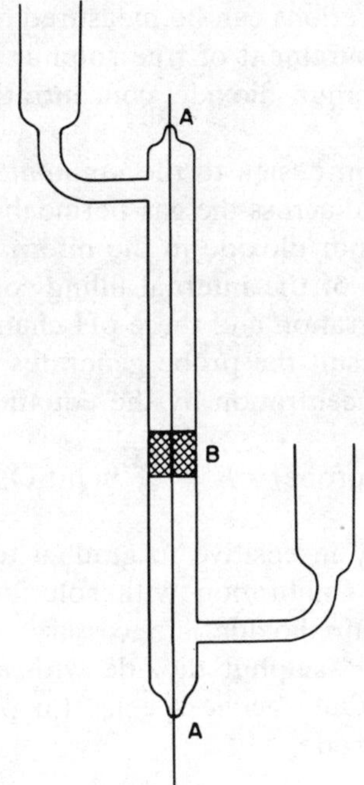


Figure S.2 Streaming potential cell: A, A, platinum wire leads to platinum electrodes; B, plug of material under study

given quantitatively by the equation

$$E = \zeta \varepsilon P / 4\pi\eta\kappa$$

which is comparable with that derived for electro-osmosis, and is obtained by similar argument. In this case $P$ is the hydrostatic pressure, and $E$ the streaming potential.

The effect has provided information about the structure of the double layer and the adsorption characteristics of the surface studied.

**Sulphur dioxide probe**

The sulphur dioxide probe is an electrode that measures the partial pressure of sulphur dioxide in solutions; sulphite, bisulphite and metabisulphite concentrations can be measured after acidification. The probe enables the measurement of free sulphur dioxide concentration and also the total sulphur dioxide concentration after preliminary alkaline treatment.

The probe is similar in design to the *ammonia probe* (q.v.); sulphur dioxide gas is transferred across the gas-permeable membrane until the partial pressure of sulphur dioxide in the internal solution equals that in the sample. The pH of the internal filling solution varies with the sulphur dioxide concentration and these pH changes are sensed by the glass electrode. As a result the probe generates a potential related to the sulphur dioxide concentration by the equation

$$E(\text{probe}) = E' + \frac{RT}{F} \ln p(SO_2)$$

The probe is virtually insensitive to gradual temperature variations in the range 0–40 °C. Calibration with solutions containing known concentrations of sulphur dioxide is necessary; the response is linear from 3000 to 3 mg $dm^{-3}$ sulphur dioxide with a non-linear response down to 0.3 mg $dm^{-3}$. Only acidic species (in particular, acetic acid) interfere with this electrode.

# T

**Tafel equation**

The Tafel equation is a relationship between the overpotential at an electrode and the current density; it can be written

$$\eta = a + b \log j$$

where $\eta$ is the *overpotential* (q.v.) expressed as a positive value, $j$ is the current density, and $a$ and $b$ are constants. This equation has been verified for a variety of reactions up to high current densities, but it is inapplicable at very low current densities where $\eta$ becomes proportional to $j$.

*See also* B & R.

**Thermodynamics of cells**
If a *reversible galvanic cell* (q.v.), of e.m.f. $E$, drives a perfectly reversible motor which delivers all the electrical energy it receives as work, then for $n$ Faradays of electricity passing through the cell and motor, the work done is $nFE$ joules. This is the maximum net work (excluding work due to change in volume) obtainable from the reaction which takes place in the cell; hence,

$$w' = -\Delta G = nFE$$

Thus, for the cell

$$\ominus\ \mathrm{Pt, H_2(g, 101\,325\ N\ m^{-2})\,|\,HCl\,|\,AgCl, Ag}\ \oplus$$

which operates on the reaction

$$\mathrm{H_2(g) + 2AgCl(s) \rightarrow 2Ag(s) + 2H^+ + 2Cl^-}$$

when 2 Faradays pass through the cell, $E(298\ \mathrm{K}) = 0.2224\ \mathrm{V}$. Hence,

$$\Delta G = -2 \times 96\,487 \times 0.2224 = -42\,917.4\ \mathrm{J}$$

If $E$ is measured at two or more temperatures, $(\partial E/\partial T)_P$ can be calculated and, hence, $\Delta H$ from the *Gibbs–Helmholtz equation*‡:

$$\Delta H = \Delta G - T\left(\frac{\partial(\Delta G)}{\partial T}\right)_P = -nF\left\{E - T\left(\frac{\partial E}{\partial T}\right)_P\right\}$$

$\Delta H$, the *enthalpy*‡ change of a reaction, can usually be determined more accurately from e.m.f. data than by direct calorimetry. The method is, however, limited to reactions which take place in a reversible cell. $\Delta S$ can be obtained from the same data; the quantity

$$nFT\left(\frac{\partial E}{\partial T}\right)_P = -T\left(\frac{\partial(\Delta G)}{\partial T}\right)_P = T\Delta S$$

is sometimes called the 'reversible heat absorbed' during the working of the cell. If $(\partial E/\partial T)$ is positive (negative), then heat is absorbed (evolved) in the working of the cell, and the electrical energy obtained is greater (less) than the decreases in the enthalpy; for the special case when $(\partial E/\partial T) = 0$, then $\Delta H = \Delta G$. The *Daniell cell* (q.v.) approximates to this latter condition.

*See also* A, De, G, J & P.

**Transference number**

The transference number, $n_x$, has been defined as the number of moles of the species $x$ transferred in the direction of the positive current for the passage of 1 Faraday of electricity.

For an electrolyte that dissociates normally, e.g.

$$LiCl \rightarrow Li^+ + Cl^-$$

the *transport number* (q.v.) of the cation is the fraction of the current carried by the cation, and

$$t_c = \frac{\Lambda(Li^+)c(Li^+)}{\Lambda(Li^+)c(Li^+) + \Lambda(Cl^-)c(Cl^-)} = \frac{\Lambda(Li^+)}{\Lambda(Li^+) + \Lambda(Cl^-)}$$

This is also the number of moles of $Li^+$ transferred to the cathode for the passage of 1 Faraday. For such electrolytes the transport number and the transference number are the same, apart from the sign convention—and this is ignored by many writers who treat the two terms as synonymous.

When more than two ions are present in a solution, the situation is different. Consider a solution of sulphuric acid which contains the species $H^+$, $SO_4^{2-}$ and $HSO_4^-$. The total quantity of electricity passed through the solution will be made up of three contributions, in the proportions $\Lambda(H^+)c(H^+)$ for the cation, and $2\Lambda(SO_4^{2-})c(SO_4^{2-})$ and $\Lambda(HSO_4^-)c(HSO_4^-)$ for the anions. The transport number definition is now inapplicable, but the changes around the electrodes enable the transference numbers to be determined. An amount $\Lambda(H^+)c(H^+)$ mole of hydrogen is transferred towards the cathode while $\Lambda(HSO_4^-)c(HSO_4^-)$ mole of hydrogen move towards the anode and

$$n(H^+) = \frac{\Lambda(H^+)c(H^+) - \Lambda(HSO_4^-)c(HSO_4^-)}{\Lambda(H^+)c(H^+) + \Lambda(HSO_4^-)c(HSO_4^-) + 2\Lambda(SO_4^{2-})c(SO_4^{2-})}$$

**Transition time**

When a constant current is passed through an electrolytic cell, an ion which is being discharged will fall in concentration in the region of the electrode (see *concentration overpotential*). If the solution is unstirred, the concentration at the electrode surface will eventually approach zero (see *limiting current density*) and this stage is marked by a sudden increase in the electrode potential. The time taken to reach this point is the transition time. For a constant current density, $j$, Fick's diffusion laws lead to the result

$$\tau^{1/2} = \frac{nF}{2j} c_0(\pi D)^{1/2}$$

where $D$ is the diffusion coefficient, so that the square root of the transition time, $\tau$, is proportional to the initial concentration, $c_0$. Use is made of this in chrono-potentiometry. $j$ must be adjusted to give transition times of the order of seconds, to avoid convective mixing, and the potential difference between the working electrode and a reference electrode is recorded on a cathode ray oscillograph. Very low concentrations may be determined in this way.

**Transport number**

The conductance of a pure electrolyte solution is (normally) the sum of two contributions, the conductance of the cations and the conductance of the (equivalent number of) anions. Transport number measurements separate these two effects and enable the individual contributions to be calculated. For this reason they have made a valuable contribution to electrolyte theory. For other properties of electrolytes, such as activities, this separation into ionic contributions is impossible; the activity coefficient of an individual ion cannot be measured, a fact which causes difficulties in the definition of pH and similar concepts.

The transport numbers, $t_c$ and $t_a$, of an electrolyte are defined as the fractions of the current carried, respectively, by the cation and anion. The calculation of these quantities from transport number measurements implies a knowledge of the nature of the ion carriers present, and this usually presents no difficulties. For a pure KCl solution, for instance, the ion carriers can confidently be stated to be $K^+$ and $Cl^-$ ions, and the value of $t_a = 0.5102$ for a 0.10 mol dm$^{-3}$ solution of KCl at 25 °C shows that the fractions of current carried by $K^+$ and $Cl^-$ are 0.4898 and 0.5102, respectively. Combined with the measured molar conductivity of the salt at this concentration, $\Lambda = 128.90\ \Omega^{-1}\ \text{cm}^2$, the molar conductivity of the $K^+$ ion in this solution is $0.4898 \times 128.90 = 63.14$ and that of $Cl^-$ is $65.76\ \Omega^{-1}\ \text{cm}^2$.

Sometimes the ionization of the electrolyte is not so straightforward.

Sulphuric acid, for instance, dissociates in two stages:

$$H_2SO_4 \rightleftharpoons H^+ + HSO_4^-$$

and

$$HSO_4^- \rightleftharpoons H^+ + SO_4^{2-}$$

Dilute solutions of the acid may therefore be expected to contain three species of ions, $H^+$, $HSO_4^-$ and $SO_4^{2-}$. Transport numbers, as defined above, are inadequate to deal with this situation, and the *transference number* (q.v.) is a more logical concept in analysing the composition of the solution. A more unusual example is hydrofluoric acid, the 'anion transport numbers' of which are shown in figure T.1. Both the values found and their great dependence on concentration show that the simple ionisation scheme

$$HF \rightleftharpoons H^+ + F^-$$

does not apply, except at very great dilutions. The rise in the apparent transport number is explained by the reaction

$$F^- + HF \rightleftharpoons HF_2^-$$

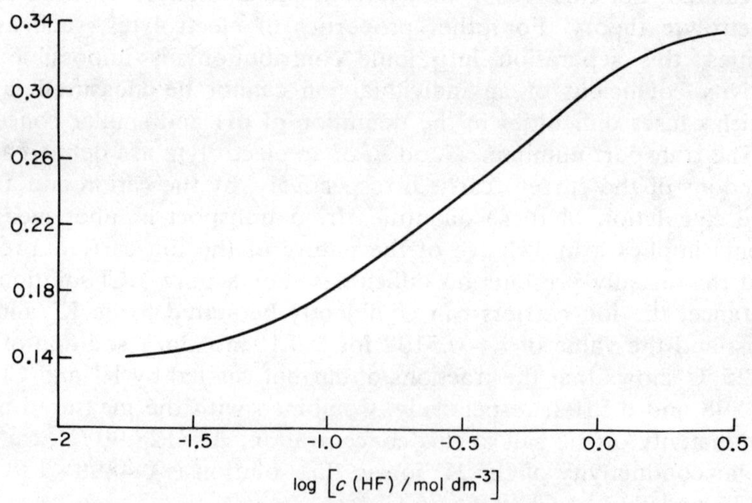


Figure T.1 Apparent anion transport numbers of HF solutions at 298 K

the complex ion carrying two equivalents of fluorine towards the anode for each Faraday of electricity passed. The same sort of situation arises whenever complex ions are formed, and transport number measurements have a valuable secondary use in detecting complex formation; they are a sensitive indication of this, because complexing not only modifies the ionic speeds, but actually causes one constituent (e.g. the H of $HF_2^-$) to move in the 'wrong' direction.

*Variation with concentration*

Transport numbers, even for electrolytes which dissociate normally, depend on the concentration. The effect is given for dilute solutions by Onsager's equation (see *conductance equations*), which gives the equivalent conductivity of an ion as

$$\Lambda_i = \Lambda_i^{\infty} - (a\Lambda_i^{\infty} + b)c^{1/2}$$

where $\Lambda_i^{\infty}$ is the equivalent conductivity at infinite dilution, $c$/equiv. dm$^{-3}$ is the concentration, and $a$ and $b$ are constants. For lithium chloride at 25 °C, for instance, this gives

$$t_c = \frac{\Lambda(\mathrm{Li}^+)}{\Lambda(\mathrm{Li}^+) + \Lambda(\mathrm{Cl}^-)} = \frac{\Lambda^{\infty}(\mathrm{Li}^+) - \{(0.2292\,\Lambda^{\infty}(\mathrm{Li}^+) + 30.16\}c^{1/2}}{\Lambda^{\infty}(\mathrm{LiCl}) - \{(0.2292\,\Lambda^{\infty}(\mathrm{LiCl}) + 60.32\}c^{1/2}}$$

Values calculated from this formula are compared with experimental results in table T.1.

Calculated and experimental values diverge more and more as the concentration becomes further removed from the very dilute range in which Onsager's equation applies.

Table T.1 Transport number of $Li^+$ in LiCl solutions of different concentrations

| Concentration /mol dm$^{-3}$ | Transport number of $Li^+$ Experimental | Calculated |
|---|---|---|
| 0.01 | 0.329 | 0.324 |
| 0.02 | 0.326 | 0.317 |
| 0.05 | 0.321 | 0.299 |
| 0.10 | 0.317 | 0.275 |

*Variation with temperature*
Ionic conductivities have large temperature coefficients, so transport numbers also vary with temperature. The general rule is that the ion with the higher conductivity has the lower temperature coefficient, so $t_c$ and $t_a$ generally become more nearly equal with rising temperature.

Accurate values of transport numbers can be obtained by the *Hittorf method* (q.v.) or by the *moving boundary method* (q.v.).

*See also* J & P, R & S.

**Tungsten electrode**
A tungsten electrode may be used to indicate the hydrogen ion concentration in solution; the *E*–pH curve for tungsten is linear over the pH range 4–9, but may be calibrated for use from pH 2–12. The *E*–pH curve for a molybdenum electrode, which behaves in a similar manner, is linear in the pH range 3–13. Both electrodes are poisoned reversibly by hydrogen and irreversibly by copper ions. They are reliable only when cleaned regularly.

# V

**Volt**
The volt, V (dimensions: $\varepsilon^{-1/2}m^{1/2}l^{1/2}t^{-1}$; units: $\mathrm{kg\,m^2\,s^{-3}\,A^{-1}}$, or $\mathrm{J\,A^{-1}\,s^{-1}}$), is the unit of electromotive force. It is the difference of potential required to make a current of 1 ampere flow through a resistance of 1 ohm.

**Voltaic cell**
*See* Reversible galvanic cell.

**Voltammetric titration**
In the voltammetric titration technique a cell is set up consisting of an electrode responsive to the substance being titrated (but not to the added reagent) and a reference electrode; the solution contains a large excess of an *indifferent electrolyte* (q.v.).

A very small current is passed during the titration. The end-point is signalled by a sudden rise in the voltage, as in a normal *potentiometric titration* (q.v.). In this case, however, it is to be interpreted as the point at which the concentration of the substance being titrated has fallen to the insignificantly low value for which the small current passing represents the limiting current (see *limiting current density*).

*See also* C, B & T, L.

**Wagner earth**
*See* Conductance, electric.

**Walden's rule**
*See* Molar ionic conductivity; Non-aqueous solutions.

**Weston cell**
The Weston cell is the most common standard cell, in which mercury forms the positive pole and a 12.5% cadmium amalgam the negative pole. The electrolyte is a saturated solution of cadmium sulphate. The cell can be formulated:

$$\ominus\quad \mathrm{Hg,\ Cd}\ \Bigg|\ \mathrm{CdSO_4\,\tfrac{8}{3}H_2O(s)}\ \Bigg|\ \begin{array}{c}\mathrm{CdSO_4\ saturated}\\ \mathrm{solution}\end{array}\ \Bigg|\ \begin{array}{c}\mathrm{Hg_2SO_4}\\ \mathrm{paste}\end{array}\ \Bigg|\ \mathrm{Hg}\quad \oplus$$

The e.m.f. of the cell is reproducible with a low temperature coefficient; at $T$ K the e.m.f. is given by

$$E(T\,\mathrm{K}) = 1.018\,30 - 4.06\times10^{-5}(T-293) - 9.5\times10^{-7}(T-293)^2$$

Standard cells must never be kept in circuit for any length of time, otherwise the e.m.f. will no longer be constant.

*See also* Potentiometer.

**Wheatstone bridge**
*See* Conductance, electric.

**Wien effect**
*See* Conductance at high field strengths.

# Z

**Zeta-potential**
*See* Electrokinetic effects.

**Zinc–air cell**
The zinc–air cell employs a zinc anode and a porous cathode of active carbon which is shaped to give a large area of exposure to the air. The electrolyte is sodium or potassium hydroxide. The reactions are similar to those of the *copper oxide cell* (q.v.), with oxygen from the air replacing CuO as the active cathode material. The electrode reactions are

$$Zn \rightarrow Zn^{2+} + 2e$$
$$\tfrac{1}{2}O_2 + 2OH^- + 2e \rightarrow H_2O + 2O^{2-}$$

the zinc ions react further:

$$Zn^{2+} + H_2O + 2O^{2-} \rightarrow OH^- + HZnO_2^-$$

and the over-all reaction is

$$Zn + \tfrac{1}{2}O_2 + OH^- \rightarrow HZnO_2^-$$

The effective voltage of the cell is about 1.2 V, and this does not fall off much during discharge.

The advantages of the cell are obvious; the disadvantage has always been that the withdrawal of a substantial current causes the cathode to become polarised, the rate of adsorption and ionisation of the oxygen being relatively slow. This drawback is steadily being reduced by improvements in the design of the cathode and in the activity of the carbon catalyst. The granules of the carbon are now surface-treated with a hydrophobic substance so that they are not wetted, and offer the maximum surface area for adsorption.

*See also* P.

**Zinc, electrometallurgy**

Zinc is prepared electroanalytically from an acid solution of purified zinc sulphate, using aluminium sheets as cathodes and anodes of pure lead. The anodic reaction is the liberation of oxygen, and so the concentration of free sulphuric acid tends to increase continuously; to counter this the electrolyte is circulated, and the more acid solutions are returned to the leaching process.

The standard electrode potential of zinc is $-0.76$ V, so it might be thought that hydrogen would be evolved preferentially at the cathode, or at least codeposited. This must be prevented, not only to maintain a high current efficiency, but because the loss of hydrogen ions from solution in contact with the cathode would result in a local alkalinity which would lead to a spongy and unsatisfactory deposit. Fortunately it is prevented by the high hydrogen *overpotential* (q.v.) at a zinc surface. A low working temperature and a high current density favour this high overpotential.

The electrolyte must be specially purified because (a) the lead anodes are attacked by an electrolyte containing more than about 50 mg $dm^{-3}$ of chloride ions and (b) almost any other metal present in the electrolyte would be deposited on the cathode, and would then tend to lower the necessary high hydrogen overpotential (see *corrosion*). The current efficiency is about 90%, an inevitable loss being the chemical attack of the zinc by the acid electrolyte.

# RECOMMENDED REFERENCE BOOKS AND TEXTBOOKS


| | |
|---|---|
| A | Adam, N. K. (1956), *Physical Chemistry*, Clarendon Press. |
| Al | Allen, M. J. (1958), *Organic Electrode Processes*, Chapman and Hall. |
| B | Britton, H. T. S. (1956), *Hydrogen Ions*, Chapman and Hall. |
| B & R | Bockris, J. O. and Reddy, A. K. N. (1970), *Modern Electrochemistry*, Vol. 2, Macdonald. |
| C, B & T | Charlot, G., Badoz-Lambling, J. and Tremillon, B. (1962), *Electrochemical Reactions*, Elsevier. |
| D | Davies, C. W. (1962), *Ion Association*, Butterworths. |
| De | Denaro, A. R. (1971), *Elementary Electrochemistry*, Butterworths. |
| E | Evans, U. R. (1960), *The Corrosion and Oxidation of Metals*, Arnold. |
| F | Falkenhagen, H. (1953), *Elektrolyte*, transl. R. P. Bell, Clarendon Press. |
| Fr | Fried, I. (1973), *The Chemistry of Electrode Processes*, Academic Press. |
| G | Glasstone, S. (1942), *Introduction to Electrochemistry*, van Nostrand. |
| H, J & S | Hughes, M. N., James, A. M. and Silvester, N. R. (1970), *S.I. Units and Conversion Tables*, Machinery Publishing Co. |
| H & O | Harned, H. S. and Owen, B. B. (1958), *The Physical Chemistry of Electrolytic Solutions*. Reinhold. |
| I & J | Ives, D. J. G. and Janz, G. J. (1961), *Reference Electrodes*, Academic Press. |
| J | James, A. M. (1976), *A Dictionary of Thermodynamics*, Macmillan. |
| J & P | James, A. M. and Prichard, F. E. (1974), *Practical Physical Chemistry*, Longman. |
| K & L | Kolthoff, I. M. and Lingane, J. J. (1952), *Polarography*, Interscience. |
| L | Lingane, J. J. (1958), *Electroanalytical Chemistry*, Interscience. |
| L & R | Lewis, G. N. and Randall, M. (1961), *Thermodynamics*, 2nd Edn, revised by Pitzer, K. S. and Brewer, L., McGraw-Hill. |
| M | McGlashan, M. L. (1968), *Physico-chemical Quantities and Units*, Monograph 15, Royal Institute of Chemistry. |
| Mi | Milazzo, G. (1963), *Electrochemistry*, transl. P. J. Mill, Elsevier. |
| Mo | Moore, W. J. (1972), *Physical Chemistry*, Longman. |
| P | Palin, G. R. (1969), *Electrochemistry for Technologists*, Pergamon Press. |
| Pa | Parsons, R. (1961), in *Advances in Electrochemistry and Electrochemical Engineering*, Vol. 1, Interscience. |
| Pu | Pungor, E. (1965), *Oscillometry and Conductometry*, Pergamon. |
| R & S | Robinson, R. A. and Stokes, R. H. (1970), *Electrolyte Solutions*, Butterworths. |
| S, G & W | Strouts, C. R. N., Gilfillan, J. H. and Wilson, A. N. (Eds.) (1962), *The Working Tools, I & II*, Oxford U.P. |
| V | Vogel, A. I. (1961), *A Textbook of Quantitative Inorganic Chemistry*, Longman. |

# TABLES OF USEFUL DATA

Table A.I. Recommended values of physical constants

| Physical constant | Symbol | Value |
|---|---|---|
| Acceleration due to gravity | $g$ | $9.81\ \mathrm{m\,s^{-2}}$ |
| Avogadro constant | $N_A$ | $6.022\,52 \times 10^{23}\ \mathrm{mol^{-1}}$ |
| Bohr magneton | $\mu_B$ | $9.273\,2 \times 10^{-24}\ \mathrm{A\,m^2 (J\,T^{-1})}$ |
| Boltzmann constant | $k$ | $1.380\,54 \times 10^{-23}\ \mathrm{J\,K^{-1}}$ |
| Charge-to-mass ratio | $e/m$ | $1.758\,796 \times 10^{11}\ \mathrm{C\,kg^{-1}}$ |
| Curie | Ci | $37.0 \times 10^{9}$ disintegrations per second |
| Electronic charge | $e$ | $1.602\,10 \times 10^{-19}\ \mathrm{C}$ |
| Faraday constant | $F$ | $9.648\,70 \times 10^{4}\ \mathrm{C\,mol^{-1}}$ |
| Gas constant | $R$ | $8.314\,3\ \mathrm{J\,K^{-1}\,mol^{-1}}$ |
| Gravitational constant | $G$ | $66.7 \times 10^{-12}\ \mathrm{m^3\,kg^{-1}\,s^{-2}}$ |
| 'Ice-point' temperature | $T_{ice}$ | $273.150\ \mathrm{K}$ |
| Molar volume of ideal gas at s.t.p. | $V_m$ | $2.241\,36 \times 10^{-2}\ \mathrm{m^3\,mol^{-1}}$ |
| Permeability of a vacuum | $\mu_0$ | $4\,\pi \times 10^{-7}\ \mathrm{kg\,m\,s^{-2}\,A^{-2} (H\,m^{-1})}$ |
| Permittivity of a vacuum | $\varepsilon_0$ | $8.854\,185 \times 10^{-12}\ \mathrm{kg^{-1}\,m^{-3}\,s^4\,A^2 (F\,m^{-1})}$ |
| Planck constant | $h$ | $6.625\,6 \times 10^{-34}\ \mathrm{J\,s}$ |
| Rydberg constant | $R_\infty$ | $1.097\,373\,1 \times 10^{7}\ \mathrm{m^{-1}}$ |
| Standard pressure, atmosphere | $P$ | $101\,325\ \mathrm{N\,m^{-2}}$ |
| Stefan–Boltzmann constant | $\sigma$ | $5.669\,7 \times 10^{-8}\ \mathrm{W\,m^{-2}\,K^{-4}}$ |
| Triple point of water | | $273.16\ \mathrm{K}$ (exactly) |
| Unified atomic mass constant | $m_u$ | $1.660\,43 \times 10^{-27}\ \mathrm{kg}$ |
| Velocity of light in a vacuum | $c$ | $2.997\,925 \times 10^{8}\ \mathrm{m\,s^{-1}}$ |
| Wien's radiation law | $\lambda_{max} \times T$ | $2.897\,8 \times 10^{-3}\ \mathrm{m\,K}$ |

Table A.II. Dielectric constants of solvents (at 20 °C unless otherwise stated)

| Solvent | Dielectric constant |
|---|---|
| *N*-methyl acetamide (40 °C) | 165 |
| Hydrocyanic acid | 115 |
| Formamide | 109 |
| Water | 80.36 |
| Formic acid | 57 |
| Ethylene glycol | 41.2 |
| Ethanolamine (25 °C) | 37.7 |
| Acetonitrile (25 °C) | 37 |
| Methanol | 33.7 |
| Benzonitrile (25 °C) | 25.2 |
| Ethanol (25 °C) | 24.5 |
| Ammonia (−33 °C) | 22.0 |
| Acetone | 21.45 |
| Cyclohexanone (25 °C) | 18.3 |
| Sulphur dioxide (0 °C) | 15.4 |
| Ethylenediamine (25 °C) | 14.2 |
| Pyridine (25 °C) | 12.3 |
| Ethylene chloride (25 °C) | 10.2 |
| Methylamine | 10.0 |
| Acetic acid | 6.1 |
| Chloroform | 4.8 |
| Diethyl ether | 4.38 |
| Benzene | 2.29 |
| Dioxan | 2.24 |

Table A.III. Standard (reduction) potentials at 298 K

| Electrode | $E^\ominus$/V | Electrode | $E^\ominus$/V |
|---|---|---|---|
| $Li^+$, Li | −3.024 | $H_3O^+$, Pt, $H_2$ | 0.000 |
| $K^+$, K | −2.924 | AgBr, Ag, $Br^-$ | 0.073 |
| $Na^+$, Na | −2.714 | AgCl, Ag, $Cl^-$ | 0.2225 |
| $Mg^{2+}$, Mg | −2.37 | Calomel (saturated) | 0.242 |
| $Al^{3+}$, Al | −1.66 | HgCl, Hg, $Cl^-$ | 0.2681 |
| $Zn^{2+}$, Zn | −0.761 | $Cu^{2+}$, Cu | 0.339 |
| $Fe^{2+}$, Fe | −0.441 | $I_2$, Pt, $I^-$ | 0.535 |
| $Cd^{2+}$, Cd | −0.402 | $Ag^+$, Ag | 0.799 |
| $PbSO_4$, Pb, $SO_4^{2-}$ | −0.350 | $Hg_2^{2+}$, Hg | 0.799 |
| $Co^{2+}$, Co | −0.283 | $Br_2$, Pt, $Br^-$ | 1.065 |
| $Ni^{2+}$, Ni | −0.236 | $Cl_2$, Pt, $Cl^-$ | 1.358 |
| AgI, Ag, $I^-$ | −0.151 | | |
| $Sn^{2+}$, Sn | −0.140 | | |
| $Pb^{2+}$, Pb | −0.126 | | |

Further values are listed in Latimer, *The Oxidation States of the Elements*, Prentice-Hall (1952).

Table A.IV. Standard redox potentials at 298 K

| Electrode system | $E^{\ominus}$/V |
|---|---|
| $Cr^{3+}$, $Cr^{2+}$ | −0.41 |
| $Tl^{3+}$, $Tl^{2+}$ | −0.37 |
| $Sn^{4+}$, $Sn^{2+}$ | 0.15 |
| $Cu^{2+}$, $Cu^{+}$ | 0.159 |
| $Fe(CN)_6^{3-}$, $Fe(CN)_6^{4-}$ | 0.356 |
| Quinone, hydroquinone, $a(H^+) = 1$ | 0.6995 |
| $Fe^{3+}$, $Fe^{2+}$ | 0.783 |
| $Hg^{2+}$, $Hg_2^{2+}$ | 0.91 |
| $Cr_2O_7^{2-}$, $Cr^{3+}$ | 1.36 |
| $MnO_4^-$, $H^+$, $Mn^{2+}$, $a(H^+) = 1$ | 1.52 |
| $Ce^{4+}$, $Ce^{3+}$ | 1.61 |
| $Co^{3+}$, $Co^{2+}$ | 1.81 |

Table A.V. pH values of aqueous solutions recommended for calibration of glass electrodes (B.S. 1647:1961)

| | pH values at 25 °C | pH values at 38 °C |
|---|---|---|
| 0.1 mol dm$^{-3}$ potassium tetroxalate | 1.48 | 1.50 |
| 0.1 mol dm$^{-3}$ hydrochloric acid and 0.09 mol dm$^{-3}$ potassium chloride | 2.07 | 2.08 |
| 0.05 mol dm$^{-3}$ potassium hydrogen phthalate (primary standard) | 4.005 | 4.026 |
| 0.1 mol dm$^{-3}$ acetic acid and 0.1 mol dm$^{-3}$ sodium acetate* | 4.64 | 4.65 |
| 0.01 mol dm$^{-3}$ acetic acid and 0.01 mol dm$^{-3}$ sodium acetate* | 4.70 | 4.72 |
| 0.025 mol dm$^{-3}$ disodium hydrogen phosphate and 0.025 mol dm$^{-3}$ potassium dihydrogen phosphate | 6.85 | 6.84 |
| 0.05 mol dm$^{-3}$ sodium tetraborate | 9.18 | 9.07 |
| 0.025 mol dm$^{-3}$ sodium bicarbonate and 0.025 mol dm$^{-3}$ sodium carbonate | 10.00 | |

* Prepared from pure acetic acid, diluted and half neutralised with sodium hydroxide. It must not be prepared from sodium acetate.

Table A.VI. Colour changes and pH range of some acid–base indicators

| Indicator | Colour: Acid | Colour: Alkaline | Approx. pH range | $pK_{In}$ |
|---|---|---|---|---|
| Methyl violet | yellow | violet | 0.1–2.0 | |
| Thymol blue | red | yellow | 1.2–2.8 | 1.7 |
| Bromo-phenol blue | yellow | blue | 2.9–4.6 | 4.0 |
| Methyl orange | red | yellow | 3.1–4.4 | 3.7 |
| Methyl red | red | yellow | 4.2–6.3 | 5.1 |
| Bromo-cresol purple | yellow | violet | 5.2–6.8 | 6.3 |
| Bromo-thymol blue | yellow | blue | 6.0–7.6 | 6.3 |
| 4-Nitrophenol | colourless | yellow | 5.6–7.6 | 7.1 |
| Phenol red | yellow | red | 6.8–8.4 | 7.9 |
| Thymol blue | yellow | blue | 8.0–9.6 | 8.9 |
| Phenolphthalein | colourless | red | 8.3–10.0 | 9.6 |
| Thymolphthalein | colourless | blue | 8.3–10.5 | 9.2 |
| Alizarin yellow R | yellow | red | 10.0–12.0 | |
| Tropaeolin O | yellow | orange | 11.1–12.7 | |

Table A.VII. Colour changes for some oxidation–reduction indicators

| Indicator | Colour: Oxidised form | Colour: Reduced form | $E^{\ominus}/V$ (pH = 0) |
|---|---|---|---|
| Pheno-safranine | red | colourless | 0.28 |
| Methylene blue | blue | colourless | 0.52 |
| Diphenylamine | violet | colourless | 0.76 |
| Diphenylamine–sulphonic acid | red-violet | colourless | 0.85 |
| Lissamine green | orange | green | 0.99 |
| *N*-phenylanthranilic acid | purple-red | colourless | 1.08 |
| *o*-Phenanthroline | blue | red | 1.08 |